T0121236

Understanding Reproduction

Our understanding of reproduction and reproductive processes is often biased towards the behaviour of organisms most familiar to us. As a consequence, the amazing diversity in the phenomena of reproduction and sex is often overlooked.

Understanding Reproduction addresses all the main facets of this large chapter of the life sciences, including discussions of asexual reproduction, parthenogenesis, sex determination, reproductive effort and much more. The book features an abundance of examples from across the tree of life, including animals, plants, fungi, protists and bacteria.

Written in an accessible and easy-to-digest style, overcoming the intimidating diversity of the technical terminology, this book will appeal to interested general readers, biologists, science educators, philosophers and medical doctors.

Giuseppe Fusco is Associate Professor of Zoology in the Department of Biology of the University of Padova. His research is in the area of evolutionary biology, with a focus on the variation produced in each generation through reproduction and development, the 'raw material' on which natural selection and other mechanisms of evolutionary change operate. He is editor of the volumes *Evolving Pathways: Key Themes in Evolutionary Developmental Biology* (2008), *From Polyphenism to Complex Metazoan Life Cycles* (2010), *Arthropod Biology and Evolution: Molecules, Development, Morphology* (2013), *Perspectives on Evolutionary and Developmental Biology* (2019) and author, with Alessandro Minelli, of *The Biology of Reproduction* (2019).

Alessandro Minelli was Professor of Zoology at the University of Padova until his retirement in 2011. He previously served as the Speciality Chief Editor for evolutionary developmental biology for the journal *Frontiers in Ecology and Evolutionary Biology*. He was previously Vice-President of the European Society for Evolutionary Biology. Having studied animals for the majority of his career, on retirement he decided to study plant evolutionary development.

He is the author of *Biological Systematics* (1993), *The Development of Animal Form* (2003), *Forms of Becoming* (2009), *Perspectives in Animal Phylogeny and Evolution* (2009), *Plant Evolutionary Developmental Biology* (2018), *The Biology of Reproduction* (2019, with Giuseppe Fusco) and *Understanding Development* (2021).

The **Understanding Life** series is for anyone wanting an engaging and concise way into a key biological topic. Offering a multidisciplinary perspective, these accessible guides address common misconceptions and misunderstandings in a thoughtful way to help stimulate debate and encourage a more in-depth understanding. Written by leading thinkers in each field, these books are for anyone wanting an expert overview that will enable clearer thinking on each topic.

Series Editor: Kostas Kampourakis http://kampourakis.com

Published titles:

Understanding Reproduction

GIUSEPPE FUSCO
University of Padova

ALESSANDRO MINELLI
University of Padova

CAMBRIDGE
UNIVERSITY PRESS

Shaftesbury Road, Cambridge CB2 8EA, United Kingdom

One Liberty Plaza, 20th Floor, New York, NY 10006, USA

477 Williamstown Road, Port Melbourne, VIC 3207, Australia

314–321, 3rd Floor, Plot 3, Splendor Forum, Jasola District Centre,
New Delhi – 110025, India

103 Penang Road, #05–06/07, Visioncrest Commercial, Singapore 238467

Cambridge University Press is part of Cambridge University Press & Assessment,
a department of the University of Cambridge.

We share the University's mission to contribute to society through the pursuit of
education, learning and research at the highest international levels of excellence.

www.cambridge.org
Information on this title: www.cambridge.org/9781009225939

DOI: 10.1017/9781009225922

© Giuseppe Fusco and Alessandro Minelli 2023

This publication is in copyright. Subject to statutory exception and to the provisions
of relevant collective licensing agreements, no reproduction of any part may take place without
the written permission of Cambridge University Press & Assessment.

First published 2023

A catalogue record for this publication is available from the British Library.

A Cataloging-in-Publication data record for this book is available from the Library of Congress.

ISBN 978-1-009-22593-9 Paperback

Cambridge University Press & Assessment has no responsibility for the persistence
or accuracy of URLs for external or third-party internet websites referred to in this publication and
does not guarantee that any content on such websites is, or will remain, accurate or appropriate.

'Fusco and Minelli provide a very clear and accessible overview of the strange and wonderful diversity of reproductive strategies and mechanisms in animals, plants and other organisms. They explain key concepts, define important terms, and place reproductive modes within an ecological and evolutionary context. This book will be a useful reference for biologists, students and even curious non-specialists.'

Russell Bonduriansky, University of New South Wales, Australia

'As a plant biologist, I often find myself trying to explain reproduction in plants as though they are somehow an anomaly rather than just another way of reaching the same goal following first principles. This perception of anomaly comes from a pedagogical bias of teaching reproduction as "sex in mammals". This book ties together concepts regardless of organism, drawing clear lines between a complex diversity of patterns and their underlying reproductive processes.'

Chelsea D. Specht, Barbara McClintock Professor in Plant Biology, Cornell University, USA

Contents

Foreword

Everybody has heard of reproduction; it has to do with sex and sexes, but not always. When it comes to humans, it is at the same time both an interesting and a taboo topic. Teachers may find it easier to teach about reproduction in plants rather than animals, and students may be puzzled when they are told that the beautiful flowers they have in that vase in their living room are in fact reproductive organs – I do not think that anyone would like the idea of having the reproductive organs of animals exposed in a vase in the living room! But this also shows the biased way in which we think about reproduction, taking ourselves and other animals – particularly mammals – as the starting point of any such discussion. But as Giuseppe Fusco and Alessandro Minelli show in this marvellous and informative book, there is a lot more to sex and reproduction than what we are usually familiar with. The present book is a real journey during which the authors masterfully describe and explain the diversity of forms of sex and reproduction across the whole tree of life. Reading this book will make you perceive sex and reproduction in a new way, far removed from the anthropocentricism that usually characterizes our perceptions of the subject. One message of the book is that we cannot draw on nature in order to make decisions about what is normal or abnormal when it comes to human sex and reproduction. What happens regularly in some groups may never happen in others; what could be considered as ordinary in some cases might be exceptional for others. Life has evolved a variety of ways for reproduction to occur, and Fusco and Minelli are to be commended for bringing us such a rich view of this complex topic in a single concise volume.

Kostas Kampourakis, Series Editor

Preface

We all have an intuitive idea of what reproduction is and how it occurs in the living world. Images that might come to mind are those of a lioness licking her cubs, a puffball mushroom that explodes releasing spores, two damselflies copulating on a reed leaf, a bumble bee that brings the pollen collected from the stamens of a snapdragon to the pistils of another and possibly many others. All of this is right, but there is so much more than that. Common understanding of reproduction and reproductive processes is biased towards the behaviour of organisms (especially animals) more familiar to us. At variance with other fields of biology such as development or evolution, in this area of biology there are no widespread 'misconceptions', but rather a lack of appreciation of the amazing diversity of the phenomena of reproduction and sex. These are often unexpectedly different from what everybody knows from humans or other vertebrates. The variation lies not only in the way parents provide care for offspring, but in whether or not they do; not only in how males court females, but in who courts whom. There is variation in the number of sexes, the number of parents, what eggs do with sperm, whether offspring are the same kind of animal, or plant, as their parents.

Inspired by a graduate textbook we published in 2019, *The Biology of Reproduction*, to which this book owes most of its field-specific contents, we will address all the main facets of this large chapter of the life sciences – asexual reproduction, parthenogenesis, sex determination, reproductive effort and much more – with a coverage across the tree of life that extends across all the main groups: animals, plants (including 'algae'), fungi, protists and bacteria. While dealing with subjects as varied as the binary division of unicellular algae, the splitting of a strawberry's stolon from the mother plant, the mating of squid, the production of spores by boletus mushrooms, or the paternal care of Darwin's frog, we will try to present all these phenomena using a common language for all living beings, at least for the most general aspects of their reproductive biology, thus overcoming the diversity of the technical terminology still alive in different disciplinary traditions, such as in botany, zoology, microbiology and transmission genetics.

This is not a book of curiosities. How do porcupines do it? (in some way). How long is a baleen whale's penis? (very long). How many times can a lion copulate in a single day? (several). What we propose here is a planned journey through what we may call a 'phenomenology of reproduction'. The first chapter will prepare the ground with some preliminary concepts. There, we will also explain the difference between investigating sex and reproduction across the whole tree of life and in relation to our own species, where psychology and social sciences get involved. After an exploration of the great diversity of life cycles (Chapter 2), we will start with reproduction that does not involve sex (Chapter 3), before moving on in Chapter 4 to sexual reproduction, then to sexual reproduction in its most canonical form, where the parents are two (Chapter 5), and then to other, less well-known forms of sexual reproduction, where the parent is one, or nearly so (Chapter 6). In the last two chapters we will have a look at how sexual traits develop (Chapter 7), before closing with the broad subject of reproductive strategies (Chapter 8).

We will not systematically discuss the evolution and the adaptive value of different reproductive modes or strategies, but our classification of reproductive phenomena largely reflects the way these are relevant in evolutionary ecology and evolutionary developmental biology. During the journey we will encounter, along with more 'regular' reproductive modes, offspring produced solely for the purpose of nourishing their siblings, plants that entrust their pollen to bats, animals that have not known sex for tens of millions of years, insects with very many X and Y sex chromosomes, apparently immortal trees and bugs that are generated by mothers that are still to be delivered by their own mother. In the very last part of the book, we will see cases where there are doubts on how many individuals, either parents or offspring, have to be counted and cases at the border of what we can call reproduction, bearing a closer resemblance to the development or the growth of an individual.

Let's start our exploration of this overwhelming variety of forms through which reproduction has evolved along the larger and smaller branches of the great tree of life.

Acknowledgements

We are grateful to Katrina Halliday and Kostas Kampourakis for inviting us to write this book.

Along the way to publication, we enjoyed the precious assistance of Olivia Boult, Kostas Kampourakis, Jessica Papworth, Jenny van der Meijden and Hugh Brazier.

Several colleagues and friends have contributed significantly to the genesis (or generation) of this work, either helping with our previous book, *The Biology of Reproduction* (2019), to which the contents of this book owe so much, or directly commenting, providing information or help on specific issues in this one, or both: Wallace Arthur, Loriano Ballarin, Giorgio Bertorelle, Ferdinando Boero, Roberto Carrer, Maurizio Casiraghi, James DiFrisco, Andrea Di Nisio, Diego Fontaneto, Carlo Foresta, Cora Fusco, Clelia Gasparini, Adriana Giangrande, Diego Hojsgaard, Marta Mariotti, Koen Martens, Francesco Nazzi, Pietro Omodeo, Marco Passamonti, Andrea Pilastro, Valerio Scali, Emanuele Serrelli, Irene Stefanini, Antonio Todaro and Vera Tripodi.

Mariagiulia Sottoriva produced all the final artwork, and significantly contributed to designing its adaptation for the present book from more technical sources. We thank her for her patience in dealing with our not always clear ideas on what the final artwork should be.

1 Individuals and Reproduction

Forms of Persistence

Ever since living beings arose from non-living organic compounds on a primordial planet, more than 3.5 billion years ago, a multitude of organisms has unceasingly flourished by means of the reproduction of pre-existing organisms. Through reproduction, living beings generate other material systems that to some extent are of the same kind as themselves. The succession of generations through reproduction is an essential element of the continuity of life. Not surprisingly, the ability to reproduce is acknowledged as one of the most important properties to characterize living systems. But let's step back and put reproduction in a wider context, the endurance of material systems.

Compared to material systems belonging to the domain of inanimate matter, such as rocks and minerals, living beings are material systems with a relatively modest degree of physical persistence: their existence depends on some capacity for 'renewal' through time. This renewal occurs at different levels of their organization and at different time scales with respect to the lifetime of the individual organism.

At the level of their most basic components, all living beings are subject to a considerable flow of matter through growth and metabolism, which has profound effects on their constitution at the molecular level. At each breath, about 10^{21} oxygen atoms, which for a while have been part of our body, abandon us in the form of CO_2 and H_2O molecules, while as many new oxygen atoms, in form of O_2 molecules from the air, become part of us. These 'new' atoms have certainly belonged to countless other living beings – microorganisms, plants, dinosaurs, bugs and multitudes of past and present humans, including ourselves some time ago. There are so many atoms even in a small chunk of matter, that this kind of atomic promiscuity is a statistical certainty. It has been calculated that more than 90% of our atoms are replaced every year. At any time, our bodies take in new atoms from the air we breathe, the food we eat, and the liquids we drink. These atoms are incorporated into our cells and fuel the chemical processes that keep us alive. This happens in all living beings, although to a variable extent, depending on the species and the life stage. For example, there is very little molecular turnover in the resting spores of many bacteria, which can remain dormant for thousands of years.

At a higher level of their organization, all living beings renew whole parts of their body during life. Unicellular organisms can renew subcellular structures, and, more prominently, multicellular organisms can renew entire cells. The turnover rate of our cells (calculated for an adult human male of 170 centimetres and 70 kilograms) varies among tissues, from 0 (cells never replaced, like most brain neurons) to 210 billion cells per day (red blood cells), for a total of 330 billion cells per day, corresponding to about 80 g of mass. About 86% of these cells are blood cells, and most of the remainder belong to the wall of the stomach and the gut. Our red blood cells persist in circulation for only about 3–4 months in an adult individual, before being phagocytized by other circulating cells, so they must be continuously replenished through cell proliferation and differentiation. The epithelial cells of our intestine are renewed every 4–5 days, so we renew more than 30 g of our gut (about 40 billion cells) per day. This kind of renewal at the cellular level is even more conspicuous in case of injuries to the body that are repaired through healing or regeneration, as when a lizard loses its tail.

Finally, since living beings are mortal (a topic we will discuss in detail before the end of this chapter), there is a population-level renewal, through reproduction. The plants that cover the slopes of a mountain are probably not the same individuals that covered those slopes a thousand years ago. A pine tree, before dying and disappearing as an individual material system, can generate other pines, and the pine forest endures, at least for a while, certainly for much longer than the time the single pine tree persists. On the longest time scale, living systems compensate by means of reproduction for their limited capacity for individual persistence.

All these forms of renewal in living beings contrast with the apparently unchanging shape of the mountain on which our pine forest sits. On the time scale of the life of the forest, the mountain has largely maintained its constitution down to the single atoms, most of which are the same and have remained in the same spatial relationships, the latter affected only to a negligible degree by movements of blocks of rock along fault lines and the surface phenomena of erosion and transport.

An exploration of this third way of living renewal–the reproductive processes – is the subject of this book, while the objective of this chapter is to equip ourselves with some conceptual tools that will be needed along our journey. To help the reader manage the diversity of organisms we will mention in the following pages, a quick guide to the diversity of life is provided in Box 1.1, while some basic biological knowledge needed to appreciate the diversity of reproductive process can be found in the last section of this chapter.

Box 1.1 A Classification of Living Organisms

To help readers move through the variegated landscape of our broad-spectrum taxonomic treatment, we outline here a classification scheme that is as up to date as possible. For convenience of exposition, we have nevertheless saved a few old names that have disappeared from current classifications because the groups they identify have been acknowledged not to be monophyletic, that is, to include all and only the descendants of a common ancestor. For example, while vertebrates are confirmed as a monophyletic group, invertebrates are simply all animals that are not vertebrates. Other commonly used names that do not correspond to monophyletic taxa are protists (unicellular eukaryotes), polychaetes (annelids other than oligochaetes, the latter including earthworms and leeches), crustaceans (in the traditional sense that excludes insects), algae, bryophytes, pteridophytes and even reptiles in the usual sense that excludes birds.

The primary taxonomic division of living beings is between prokaryotes and eukaryotes. Prokaryotes, most of which are unicellular, include the true bacteria, within which are classified also the blue-green algae, and a diverse group known as the archaea. Unicellular organisms also account for a huge and diverse set of eukaryotes, only a few of which have close affinities to multicellular ones, such as *Volvox* or *Chlamydomonas* with the true plants, or diatoms with the brown algae.

Most unicellular eukaryotes, and a few multicellular ones that are closely related to some of them, are informally called protists. Free-living aquatic photosynthesizing protists are often called algae, an informal group that includes also aquatic photosynthesizing multicellular organisms; based on their pigments, red, brown and green algae are distinguished.

The most familiar groups of eukaryotes are the land plants, the fungi and the animals.

The simplest kind of land plants (or embryophytes) are mosses and liverworts, traditionally grouped as bryophytes; they lack the vessels for the transport of sap that characterize the remaining land plants, therefore called vascular plants or tracheophytes. These comprise, in turn, horsetails and ferns, collectively called the pteridophytes, and a much larger group, the seed plants or spermatophytes. Some of the latter, the gymnosperms, including conifers, lack true flowers and produce naked ovules; their sister group are the flowering plants, or angiosperms. Here, ovules mature within the ovary, and thus the seeds are enclosed in a fruit.

Long classified with plants, but closer instead to animals, are the fungi. These include a modest number of unicellular forms (yeasts) and a huge diversity of multicellular organisms. Devoid of the pigments that characterize algae and other plants, fungi are mostly exploiters of decaying matter, or symbionts associated with the roots of plants, or parasites. The main fungal taxa are ascomycetes, basidiomycetes and zygomycetes.

Within animals (or metazoans), some 30 highest-level groups or phyla are currently accepted. Sponges are mainly marine, but a few live in freshwater. Other marine groups are the diaphanous ctenophores or comb-jellies, and nearly all cnidarians (sea anemones, corals, medusae). Many free-living members of the platyhelminths or flatworms are also marine, but the group includes freshwater forms such as planarians. Flatworms include also two large groups of parasites, the digeneans or flukes and the cestodes or tapeworms.

A generally elongated body articulated into segments is the most obvious feature of annelids. These include terrestrial or freshwater groups like earthworms and leeches, but the majority is represented by the marine polychaetes. Other animal groups are often described as minor, because they include only a handful of species or because these are small and inconspicuous, or for both reasons, but we will nonetheless encounter them in our pages because of peculiarities of their reproductive biology.

Huge groups of invertebrates are the nematodes and the arthropods. Nematodes or roundworms include very many parasites, including *Ascaris*, but also free-living species such as *Caenorhabditis elegans*, one of the most popular model organisms studied in the laboratory. Arthropods, unmistakable for their segmented body provided with an exoskeleton and more or less numerous pairs of articulated appendages, including sensory (antennae), feeding (maxillae, mandibles) and locomotory (legs), are the largest of all animal phyla. They include the chelicerates, with scorpions and pseudoscorpions, mites and spiders; the many-legged myriapods; and the crustaceans, mainly aquatic, but with a successful branch of terrestrial forms (woodlice or land isopods). Other than the familiar decapods (crabs, shrimps, lobsters), crustaceans include many other groups such as cladocerans (*Daphnia*). Based on molecular evidence, insects are classified today by most authors as a specialized lineage of crustaceans.

Very unusual and diverse are the shapes of the echinoderms (e.g. starfishes and sea urchins), and those of tunicates, the sessile ascidians and the planktonic thaliaceans. Despite the obvious morphological distance, these latter groups of marine animals are closely related to vertebrates.

Within vertebrates, the old class of fishes has been long abandoned, their content being now distributed between several major groups: we will deal with representatives of the cartilaginous fishes (sharks, rays), crossopterygians (coelacanth) and numerous bony fishes. Of amphibians, we will mention both urodels (salamanders) and anurans (frogs and toads). Of reptiles other than birds, we will see the lineages of turtles, squamates (lizards, snakes) and crocodiles.

Following a simplified but up-to-date classification, the mammals mentioned in the text belong to the following groups: monotremes (platypus and echidnas), marsupials (kangaroo, wallaby, opossum) and placentals, to which we belong.

What Is Reproduction?

Let's adopt an informal common-sense concept of reproduction. In biology, we can define reproduction as *the process by which new individuals are generated from pre-existing individuals*. Implicitly, it is assumed that the 'new individuals' are materially derived from pre-existing ones, or parents. This concept of reproduction has deep roots in human history, as it emerged by obtaining increasingly detailed knowledge of the life cycles of the plants and animals most familiar to us, ourselves included. This simple definition, as trivial as it may seem, nonetheless allows us to leave out other possible forms of generation that are not found in living beings.

A long tradition that remained alive until the nineteenth century accepted the possibility of *spontaneous generation*, that is, the direct formation of living organisms (the simplest ones at least) from inanimate matter, such as mud or rotting organic material. It was William Harvey (1578–1657), who in his *Exercitationes de generatione animalium* of 1651 first recognized a fundamental divide between the world of living beings and the rest of nature, based on a principle of generation: '*omnia ex ovo*', all living beings are born of an egg. Belief in spontaneous generation was progressively dismantled through the experimental work of scientists, among them Francesco Redi (1626–1697), who inflicted a mortal blow to the doctrine of spontaneous generation with his experiments designed to reveal whether 'any exudate from a rotten corpse, or any filth of something putrefied, generates worms'. Redi observed indeed that 'worms' (actually, insect larvae) developed in sealed glass flasks containing rotting pieces of meat or little dead animals, but only if, before sealing them, access to these flasks and their content was allowed to big flies, blue, green or grey, similar to those into which the 'worms' eventually transformed. Further important contributions were provided by Lazzaro Spallanzani (1729–1799), Franz Schulze (1815–1873) and finally Louis Pasteur (1822–1895), who definitively demonstrated that spontaneous generation does not occur, not even in microorganisms.

Material systems that are not considered living organisms, or do not qualify as living cellular systems – for instance, viruses, prion proteins (prions) and transposable DNA sequences (transposons) – can reproduce in some sense, although differently from living organisms. An important difference with respect to organismal reproduction is that these systems can replicate without contributing constitutive matter to their descendants. In other words, the causal link between 'parents' and 'offspring' does not pass through the transformation of a part of the body of the parent into the offspring. For instance, prions are proteins with a special conformation, capable of inducing other molecules of the same protein, but with a different three-dimensional structure, to take their own conformation. In this way, prions start a chain reaction that leads to their multiplication in an infectious way. Some prions are associated with pathological conditions of the mammal nervous system – for instance, bovine spongiform encephalopathy (BSE), also known as mad cow disease. Another example is offered by retroviruses, such as human immunodeficiency virus (HIV). When infecting a cell, the virus induces it to produce copies of its genetic material and the molecules for its shell, the capsid. However, neither the genetic material nor the proteins of the capsid of the 'offspring generation' are derived from parental molecules. The parent virus is causally responsible for the production of a new viral unit, but does not contribute any of the matter of which it is made.

Apart from these other forms of reproduction, two different aspects of population renewal contribute to biological reproduction. One is that 'new individuals' are added to the set of existing ones. In this *demographic* aspect of reproduction, 'new' should be understood as a quantitative addition to the entities that already exist. The other is the production of 'individuals that are new', compared to the existing ones. In this *innovative* aspect of reproduction, 'new' should be understood as qualitatively different from the pre-existing entities. According to the demographic concept, focus is on replacing the individuals that inevitably perish, thus maintaining or possibly increasing the size of the population. In contrast, according to the innovative concept, focus is on the appearance of 'something new under the sun', thus allowing the population to change through the generations – that is, to evolve.

These two distinct aspects of reproduction are found together in some forms of reproduction, for instance, in the most common instances of sexual reproduction, but not in all. As we will see, some modes of reproduction lead to a renewal only in the demographic sense, while innovation, although eventually occurring in all organisms, is not necessarily associated with reproduction.

Asexual and Sexual Reproduction

Traditionally, a primary distinction is traced between *asexual* and *sexual* reproduction. This division, as the two terms suggest, is based on the characterization and involvement of the so-called *sexual processes*.

At the genetic level, *sexual processes* are biological processes through which new combinations of genetic material are created from distinct sources. These can occur through the union of genetic material of different provenance, as in the fusion of the gametes of two individuals, but also through the reassortment of genetic material of different origin, like the reshuffling of the two sets of chromosomes during meiosis (*recombination*) for producing gametes or spores. Novel genetic assets can be obtained also in other ways. One is *horizontal gene transfer*, where genetic material is moved from one individual to another that is not its offspring (and not even necessarily of the same species), to be incorporated in the latter's genome. This sexual process occurs regularly in prokaryotes, but it is common among eukaryotes and between prokaryotes and eukaryotes as well. Sexual processes, sometimes dubbed simply *sex*, for short, are thus distinguished, in principle at least, from the processes of change of genetic information that occur in a single individual without the contribution of DNA from other sources, as in gene mutations. In the case of eukaryotes, for simplicity, in this book we will discuss the genetic aspects of reproduction at the level of the nuclear genome, overlooking the dynamics internal to the genomes of the organelles (mitochondria and plastids).

These phenomena of genetic reassortment may or may not be associated with reproduction. A classic example of sex dissociated from reproduction is offered by ciliates, a group of unicellular protists to which paramecia belong. Ciliates commonly practise a form of sex called *conjugation*. Two individuals (conjugants) unite temporarily, exchange genetic material through a sort of reciprocal fertilization, then separate again. The result of this exchange is a pair of independent individuals (ex-conjugants), genetically different from the conjugants. Note that the number of individuals has not changed through sex: two conjugants in, two ex-conjugants out.

Thus, in ciliates it is possible to have reproduction without sex, when a paramecium simply splits into two (binary fission, Chapter 3), or sex without reproduction, when two paramecia unite, exchange genetic material and then separate, both genetically renovated (conjugation).

Sexual processes, associated or not with reproduction, are found in virtually all major lineages of living beings. Note that in the scientific literature about the genetics of reproduction, 'to have sex' generally means 'to carry out a sexual process', not 'to mate'. That a mating may not result in reproduction is probably quite obvious to everybody.

Reproduction can thus be qualified as sexual or asexual based on the involvement of potentially independent sexual processes. Accordingly, sexual or asexual are in principle attributes of reproduction, but not attributes that apply to a given organism, since many organisms can reproduce in both ways.

Asexual reproduction (in plants, also called *vegetative reproduction*) is a mode of *uniparental reproduction* (i.e. reproduction from a single parent) that does not involve sexual processes or the production of gametes, not even in derived or residual form. As a first approximation, asexual reproduction generates individuals genetically identical to each other and identical to the parent, thus forming what is termed a clone (Chapter 3), and for this reason it is also called clonal reproduction. However, asexual reproduction in some cases fails to produce perfectly clonal descendants, while sexual reproduction occasionally has a clonal outcome.

Asexual reproduction is the most common form of reproduction among unicellular organisms, prokaryotes and eukaryotes alike, but is also very common among multicellular organisms. For instance, many plants, for example most reeds, propagate by runners; many marine worms multiply by splitting into pieces; and some sea anemones generate new polyps through budding. In bacteria and some protists, this is the only mode of reproduction (*obligate asexual reproduction*). However, more often, asexual reproduction coexists with sexual reproduction, either as the only form of reproduction in a specific phase of an organism's life cycle (e.g. in the polyp phase of cnidarians with a typical metagenetic cycle; Chapter 2), or as a reproductive option (*facultative asexual reproduction*) co-occurring with the sexual one (e.g. in many plants). In any case, the obligate asexual reproduction of certain organisms does not rule out the possibility of these having sex that is not associated with reproduction, as mentioned above for the paramecium, and this is very common in prokaryotes.

Sexual reproduction is a form of reproduction that generates new individuals with a genetic make-up resulting from the association and/or the reassortment of genetic material of different origins. In the most canonical form of sexual reproduction, the genome of an offspring, in other words the totality of its genetic material, derives from the union of (partial) copies of the genomes of two parents through fertilization. Here two distinct sexual processes are involved: the recombination of each parent's genomes at meiosis during the production of gametes, and the combination of the genomes of the two gametes into the zygote's genome at syngamy. Sexual reproduction involving two parents is called *amphigony* or *biparental sexual reproduction*. However, there are also forms of *uniparental sexual reproduction* (Chapter 6), in which the genome of a single parent is modified and reorganized during the process that leads to the offspring generation, as in parthenogenesis (reproduction through eggs that do not need to be fertilized) or in self-fertilization (reproduction through eggs that are fertilized by the sperm of the same individual). Sexual reproduction is found in the vast majority of multicellular eukaryotes and in most protists, but it does not necessarily rule out other forms of sex or other forms of reproduction.

With reference to the demographic and innovative aspects of reproduction mentioned in the previous section, the demarcation between sexual and asexual reproduction we have just established entails both the possibility of innovation without reproduction and without demographic growth (through sex alone), and the possibility of reproduction with or without sex, and with or without the production of genetic novelty.

Unfortunately, a discordant terminology is in use, based on different opinions on where to draw the line between sexual and asexual reproduction. In particular, what is contentious is whether some forms of uniparental reproduction, such as parthenogenesis, should be considered as sexual or asexual, along with the species or the individuals that practise them. In this book we have adopted a terminology that takes into account both the role of sex and the processes by which gametes are formed. Cases of uniparental reproduction involving sexual processes (as in self-fertilization), or deriving from processes typical of sexual reproduction (such as egg formation in parthenogenesis), are treated in the context of sexual reproduction, even when they have a clonal outcome. However, this terminological inconsistency in the literature is a minor point in comparison to the fact that there are forms of reproduction that challenge inclusion in any rigid classification. In any case, the reader is alerted: sexual and asexual may mean different things to different people.

Generations

We, the authors of this book, both belong to the 'Boomer Generation', people born in the two decades following World War II. On the Internet, this label is often used in contexts where teenagers and young adults (who evidently belong to another generation) tease attitudes typical of older adults. In the Western countries, the Boomer Generation was followed by Generation X, those born between the 1960s and the 1980s, then by the Millennials (Gen Y, born from early 1980s to mid-1990s), the Zoomers (Gen Z, born from mid-1990s to early 2010s,) and finally by the current generation Alpha. This is how sociologists and popular media subdivide generations: clusters of people born in a given period of time and, accordingly, expected to have experienced the same significant events of the society to which they belong. Most members of Gen Z are children of Gen X, not of the immediately preceding Gen Y. However, in biology (as well as in other sciences, like demography), the individuals born during the same breeding season or across the same interval of years are said to belong to the same *cohort*, while the term generation has a different meaning.

In biology, in particular in evolution and genetics studies, a *generation* is a set of individuals that come into being through a given number of reproductive events, either sexual or asexual, with reference to an individual ancestor or parent pair. Generations can thus be counted and numbered. For instance, many whale species travel in family pods that include three generations: a mother whale (parental generation), plus her daughter(s) (first offspring generation) and the young children of the latter (second offspring generation). All the descendants of an individual that reproduces several times during its life belong to the same generation. For instance, all the offspring of a female elephant that has been reproducing over multiple reproductive seasons form a single generation. Similarly, two seeds of sequoia produced by the same mother plant 2,000 years apart belong to the same generation (Figure 1.1).

Belonging to a given generation is a relative characteristic, which depends on the choice of a reference individual ancestor, but another element of relativity is added in species with overlapping generations, where mating between individuals of different ages is possible. In his book *L'ordine del tempo* (*The Order of Time*), the Italian physicist Carlo Rovelli reconstructs king Leonidas'

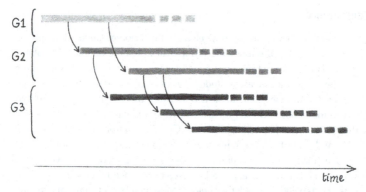

Figure 1.1. Schematic representation of the concept of generation. Horizontal thick lines are developing individuals belonging to three generations (G1–G3). Curved arrows are events of reproduction. Individuals produced at different times by different individuals, or even by the same individual, of the same (parental) generation belong to a single (offspring) generation. Note that individuals of a given generation may come into being before some individuals of a previous generation. For simplicity, a form of reproduction from a single parent is shown.

family pedigree to illustrate the relativity of time in modern physics. This is a story of 'generation relativity'.

Leonidas I, seventeenth king of Sparta, the hero of the Battle of Thermopylae during the second Persian War, married his half-brother's daughter Gorgo and with her he had a son, Pleistarchus (Figure 1.2). Pleistarchus is at the same time son (first generation) and grandson (second generation) of two brothers. Gorgo belongs to the same generation as Leonidas (as parents), but the same can be said for her father Cleomenes, Leonidas' brother (as children). With respect to Anaxandridas II, Leonidas' and Cleomenes' father, Pleistarchus belongs to the second generation through Leonidas and to the third generation through Cleomenes. Thus, generation assignment is not only relative to a given ancestor, but also relative to a specific ancestor–descendant path. The parent–offspring relationship establishes an order among some individuals belonging to a given set, but may not do so among all of them. Mathematicians use the term 'partial order' for the order provided by a relationship among the elements of a set that establishes an unambiguous precedence among some pairs of elements, but not among all pairs. Genealogy is an example of this kind of relationship.

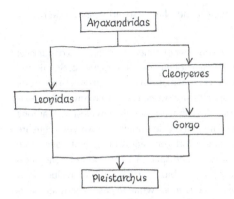

Figure 1.2. King Leonidas' pedigree. Leonidas ambiguously can be said to belong both to the same generation as Cleomenes (as a child) and to the same generation as Gorgo (as a parent). However, Gorgo does not belong to the same generation as Cleomenes, being his daughter.

This relativity in generation assignment shows up in all populations where individuals can reproduce across several breeding seasons, live long and become reproductively mature early, thus allowing overlap of generations, but the situation is even more embarrassing with some peculiar mating systems (Chapter 5). For instance, to which generation do the offspring born from the mating of a female with her own son (oedipal mating) belong? This is something that happens regularly in some organisms. Virgin females of the mite *Histiostoma murchiei* parasitize earthworm cocoons, in each of which they deposit 2–9 eggs that do not need to be fertilized. These will hatch within two or three weeks, producing only males. These males mature in about two days, mate with the mother, and die. The mother then lays about 500 fertilized eggs from which only females are born, which once developed will seek new earthworm cocoons to parasitize. These females are at the same time children and half-siblings (having the same mother) of their father, and children and grandchildren (being children of a child, their father) of the mother!

Living beings differ enormously in *generation time*, that is, in the average time interval between two consecutive generations. This interval varies from minutes for many bacteria (even as few as 12 minutes), to tens of years for large animals (20–30 years in our species) and plants (30–40 years in the beech, *Fagus sylvatica*).

Generation time has effects on the number of generations that can occur over a calendar year, or the number of years one generation may last. Small- and medium-size animals usually reach sexual maturity early and tend to have one generation per year (*univoltine* species) The same is true of many herbaceous plants (see below). This reproductive rhythm is particularly widespread among the species that inhabit regions with marked seasonality. However, there are numerous animals with particularly short generation times that have more than one generation per year (*multivoltine* species). The shortest generation time for an insect with amphigonic (i.e. biparental sexual) reproduction is a week for the mosquito *Psorophora confinnis*, while in case of parthenogenesis, generation time can be even shorter, less than five days, as in the aphid *Rhopalosiphum padi*.

In extreme cases of reproduction carried out in a juvenile stage (paedogenesis, Chapter 6), an individual can even begin to reproduce when it is still in the body of its mother. In some aphids, a parthenogenetic female carries in her body her developing daughters, and within these their own daughters (granddaughters of the former), like Russian dolls. Similar telescoped generations are found in some mites, in the stenolaematous bryozoans (marine invertebrates resembling colonial polyps), and in the colonial green alga *Volvox*.

In plants, depending on the length of the life cycle (which, however, as we will see, does not simply correspond to a single generation – see Chapter 2), we distinguish between *annual plants*, which complete the entire cycle within a year, like many herbs and grasses; *biennial plants*, which in the first year develop abundant foliage and bloom the following year, generally using the resources accumulated during the first year in bulbs, tubers or rhizomes; and *perennials*, which have longer life cycles. Annual and biennial plants are considered to have generation times of one and two years, respectively. In perennial plants the generation time depends on the age at which the plant reaches sexual maturity, which can vary from one year to several tens of years. Terrestrial plants able to complete several life cycles per year in natural conditions (corresponding to animal multivoltine species) are rare. Among these there are, in Europe, the alien populations of *Galinsoga quadriradiata*, a member of the sunflower family native to South America, which can complete 2–3 life cycles per year.

Reproduction and Individuality

Whatever way reproduction is defined, this implies accepting a suitable concept of a biological individual. Reproduction is the production of new entities, and these must correspond to some kind of individuals that can be counted. But what is an individual in biology? Rivers of ink have flowed in the attempts to answer this question, but the answer is that there is no unambiguous answer, for good biological reasons. We tend to conceive of individuals as being like ourselves, well-integrated entities, reasonably well defined in space and time, characterized by genetic homogeneity (all the cells in our body have the same genome) and genetic uniqueness (no one else has our genome, unless we have a twin), as well as by physiological unity and autonomy. However, that is simply not the case for many living beings.

Individuals lacking a *unique genome* are commonplace. Two amoebae that have just originated by fission from a parent amoeba share an almost identical genome, as do all the strawberry shoots derived from the same runner, and a pair of human identical twins. In all these cases, however, and particularly in the last one, we would be inclined to recognize these as distinct individuals. All forms of asexual reproduction and all forms of sexual reproduction with a clonal outcome undermine a definition of individuality based on genetic uniqueness. Asexual reproduction is very widespread among plants, so that botanists use two distinct terms to indicate two different kinds of 'plant individuals'. A *genet* is a genetically unique entity, either a single physiologically independent entity, or a set of entities derived from a single individual by clonal multiplication (e.g. all the strawberry plantlets derived from the same runner). A *ramet*, instead, is an anatomically and physiologically bounded biological entity, with or without genetic uniqueness (e.g. any isolated strawberry plant, either derived from seed or from a runner) (Chapter 3).

As for the *genetic uniformity* of the individual, it is highly unlikely that the nuclei of the many cells of a multicellular organism have 100% identical genomes. When referring to the genetic identity among the members of a clone (such as the cells of a multicellular organism), it is implied that the mutations accumulated in the subsequent divisions starting from the founder cell are overlooked. The condition of an individual carrying different genomes that originated from the genome of a single founder cell is called *genetic mosaicism*. This common within-organism genetic heterogeneity, normally considered negligible, may become relevant in very old or very large organisms (i.e. those with many cells), in which the last common ancestor of two cells in the same individual's body may be traced many cycles of cell division back. The most recent common ancestor of two reproductive cells of the same oak might have lived centuries ago, rather like the progenitor of a whole population of individuals in a species with an annual life cycle.

Within-organism genetic heterogeneity can also take the form of a *genetic chimerism*. A chimera (in biology, not in mythology) is a multicellular individual made of cell populations originating from more than one founder cell. Chimeric individuals can originate from the fusion of several spores in the red alga *Gracilaria chilensis*, from the fusion of larvae of the same species in freshwater sponges such as *Spongilla*, and a chimeric fungus can easily originate by fusion between hyphae of distinct individuals. But chimerism is also known among mammals. New World monkeys of the genus *Callithrix*, commonly known as marmosets, generally give birth to two (non-identical) twins. But these are not 'normal' twins. During pregnancy, connections between the two placentas are established so that cell exchanges occur between the two embryos. When the two little monkeys are born, each of them is a mixture of cells derived from the independent fertilizations of two distinct eggs. Thus, through this form of reproduction, two 'genetic individuals' and two 'physiological individuals' are obtained, but the two genetic individuals are distributed between the two physiological individuals. Furthermore, the two sperm that fertilized the two eggs can come from distinct fathers. This genetic promiscuity in parent–offspring relationships perhaps accounts for the peculiar and highly cooperative parental-care system in these monkeys. Possibly little known is the fact that a form of chimerism frequently affects our own species, together with other placental mammals. During pregnancy, fetal cells can migrate to the mother and vice versa. These cells can persist, multiply and even differentiate in the mother and the offspring for decades after birth (*fetomaternal* and *maternofetal microchimerism*).

A biological individual would be expected to be an undivided morphological and functional living unit, able to relate to the environment independently, including the ability to properly respond to environmental stimuli and the faculty to reproduce. However, individuals lacking *autonomy* and *physiological unity* are found among the members of highly integrated colonies, such as those of some marine invertebrates. Colonial hydrozoans known as the Portuguese man o' war (*Physalia*) behave as an integrated unit to the point that the colony is often mistaken for an individual jellyfish. In each colony different types of individuals (zooids) coexist and cooperate, only some of which (*gonozooids*) are able to reproduce. In *Volvox*, a colonial green alga, only the special reproductive cells called gonidia, which are located within the sphere formed by flagellate somatic cells, can reproduce asexually to form new daughter colonies. Individuals lacking reproductive autonomy also include members of the sterile castes in some animal societies, like those of many species of bees, wasps, ants, termites and, unique among mammals, the naked mole-rat (*Heterocephalus glaber*).

Colonies and advanced societies pose a problem for the interpretation of the reproductive process, because the unit of reproduction can be identified at more than one level of biological organization – for instance, either in the single zooid of a colony or in the colony as a whole, which in the latter case can be seen as a 'superorganism'.

Finally, autonomy becomes a myth when symbiosis is taken into account. Here the 'boundaries' of the individual definitely blur out. A large community of symbiotic microorganisms 'inhabits' the body of most multicellular organisms, which is necessary for their normal development and regular functioning. Our digestive system hosts a number of bacterial cells of the same order of magnitude as the cells (the 'human cells' in the strict sense) of our entire body (3.5×10^{13}). In each person, these bacterial cells belong to a few hundred different 'species', out of the thousands which have been found in symbiosis with humans as a whole.

Beyond variation in attributes, individuals in the tree of life are not the same thing in another important respect: evolution. In their landmark book *The Major Transitions in Evolution* (1995), John Maynard Smith and Eörs Szathmáry analysed and discussed the recurrent events of emergence of new levels of organization in living systems. Multiple times, across a time span of more than two billion years, a certain number of entities originally able to survive and reproduce autonomously merged into a higher-level entity, causing a new hierarchical level of biological organization to appear. Among these events are the symbiotic union of primitive prokaryotic cells in the first eukaryotic cell, the evolution of multicellularity in several lineages of eukaryotic unicellular organisms, the evolution of colonial organisms and animal societies. Thus, a bee colony is an individual made of thousands of organisms, each made of billions of eukaryotic cells, each originally derived by the fusion of two prokaryotic cells. These 'major evolutionary transitions' are actually 'evolutionary transitions in individuality'. At each one of these events, a new kind of individual has evolved through cooperation and integration of pre-existing individuals. However, these new individuals have not replaced the older ones. Rather, they have added a new level of individuality in the tree of life, increasing its diversity at this fundamental level.

Summing up, there are different kinds of individual out there. This may seem a complication for our schemes and models, but this is life, always intolerant of our simplistic definitions. However, we can cope with that. Keeping a flexible attitude towards the concept of an individual will be important in the following chapters, where we will see reproduction at work in different kinds of organism. The kind of individual and the mode of reproduction will define the boundaries between reproduction and other biological processes occurring in the same organism, like growth or development, but these boundaries are not always neat. We will highlight these 'difficult boundaries' along the way, to return with some final considerations towards the end of the book.

Reproduction and Senescence

Reproduction allows the persistence of a species despite the continual loss of its individual members. Mortality rate per unit of time in a population is never zero. Individual living beings do not last forever; they are mortal for two main reasons. First, an individual may die by an accident related to its interactions with predators or parasites, or to environmental physical factors beyond its tolerance threshold. Second, even in environmental conditions ideal for its survival, with no limit to the availability of resources for its maintenance, the individual cannot save itself from 'certain death'. This is the inescapable result of the developmental process called *senescence* (or *ageing*), manifested with an increase in the probability of death with age. Senescence is a cumulative process, occurring at different levels of body organization (from molecules to cells, tissues and organs), which progressively or abruptly, depending on the species, corrupts metabolism and body structures, producing a deterioration of the qualities of the organism that eventually leads to its death. In some cases, as in many annual plants, the squids and the Pacific salmon, senescence is triggered quickly the first and only time the organism reproduces sexually.

Senescence is the antagonist of immortality. In biology, to be immortal does not mean that an organism cannot die, a quality reserved for certain mythological figures and comic-book superheroes. All living things may die from trauma, disease, or simply being eaten by another living being. In biology, immortality, with reference to either a cell or an individual, is rather a potentiality. It is the absence or the arrest of senescence. Whereas living beings may all die, senescence does not affect all organisms in the same way. An individual or cell that does not age, or ceases to age at some point in its existence, is said to be *biologically immortal*.

Most prokaryotes and many protists do not seem to experience senescence. Also, there is no certain evidence of senescence in some plants that live for a very long time, over 4,000 years, including some conifers (*Pinus longaeva* and *Sequoiadendron giganteum*), some sponges and sea anemones, hydras, the queens of different species of social insects (bees, ants, wasps and termites), some tube-dwelling polychaetes (*Lamellibrachia*) and certain bivalves (*Arctica*). The same may apply to the black coral *Leiopathes*, for which a maximum age of 4,265 years has been estimated. All these organisms are considered potentially immortal.

For all the living beings that age, it is not enough for reproduction to generate new individuals, in addition to those already present, or to replace those that have died. Reproduction must also ensure that newborns are actually 'young' – in other words, that they have, so to speak, 'turned back the clock of senescence' so that the population 'rejuvenates' through reproduction. To generate young individuals from old ones is an imperative for the continuity of life. This is accomplished in many different ways. Sexual reproduction has this ability to rejuvenate. Through the processes that lead to the formation of gametes in multicellular organisms, for instance, the senescence timer is effectively reset to zero. The life expectancy of a fertilized egg is definitely higher than the life expectancy of the two parents from which the gametes were produced. Within certain limits, varying from species to species, it is also independent of the parents' age.

The capacity of rejuvenation seems to be an attribute also of sex in the broad sense. Ciliates reproduce only asexually, in many species by simply dividing the cell into two (binary fission, Chapter 3), but commonly practise conjugation (see above), a form of sex where two individuals (conjugants) come into contact, mix their genetic material, then separate again (as ex-conjugants), genetically different from those that conjugated, but identical to each other.

In most ciliates, the clone that originates from an ex-conjugant after separating from its partner shows a form of *clonal senescence*, consisting of a limit to the number of cell divisions it can undergo. This number varies from species to species, but also between strains of the same species. In *Tetrahymena* this limit varies between 40 and 1,500 divisions. Moreover, the clone goes through different 'maturation stages' comparable to what in a multicellular organism we might describe as developmental phases. During an initial period of 'sexual immaturity' of the clone (measured by the number of divisions since the last conjugation), individuals can only multiply asexually, but are not able to conjugate. Then follows a period of 'sexual maturity' during which they will be able again to conjugate. Ex-conjugants will emerge from this event genetically modified, but also in some way rejuvenated, with an expected number of future cell divisions equal to the maximum possible for the species or strain. Individuals that do not conjugate, instead, may continue to multiply, but will enter a phase of gradual senescence: the rate of cell divisions will gradually slow down, eventually leading to the extinction of the clone. If they conjugate during this phase of senescence, the ex-conjugants will have an expectancy of clonal propagation below the maximum value for the species.

In some cases at least, asexual reproduction generates rejuvenated offspring as well. Examples are provided by many plants with vegetative reproduction. A branch broken away from an old willow has a good chance of taking root and developing into a new young individual with a life expectancy that does not depend on the age of the parent. There are clones of quaking aspen (*Populus tremuloides*) estimated to be more than 80,000 years old (see Concluding Remarks), of the creosote bush (*Larrea tridentata*) almost 12,000 years old, and of bracken (*Pteridium aquilinum*) almost 1,500 years old.

In unicellular organisms, prokaryote and eukaryote alike, the unequal partition of damaged cell constituents among the daughter cells, even if cell division is apparently symmetrical, can rejuvenate the clone through the propagation of damage-free cells. For instance, individual cells of the yeast *Saccharomyces cerevisiae* divide asymmetrically, so that the larger cell is called the mother cell, and the smaller the daughter cell. The mother cell is subject to a form of senescence by accumulating damaged components, so that there is a limit to the number of daughter cells (about 50) it is able to produce before finally ceasing to divide. Until a certain 'age' of the mother cell (number of divisions undergone till then), the daughter cells will not receive any damaged components from the mother, and the replicative potential (number of possible divisions) of a daughter cell, which will soon act in turn as a mother cell, will be the maximum for the species. However, as the age of the mother cell increases, the daughter cells will receive increasing quantities of damaged components, thus detaching from the mother cell with a replicative potential already reduced compared to the species' maximum (Figure 1.3).

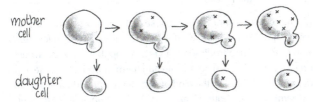

Figure 1.3. Asymmetric clonal senescence in the yeast *Saccharomyces cerevisiae*. Along the series of budding events for the same individual cell, the mother cell progressively accumulates senescence factors (crosses). Daughter cells progressively receive larger quantities of these factors. Daughter cells that are generated early retain full replication potential, while daughter cells generated later will emerge from the division of the mother cell with a reduced replication potential.

Reproduction and Us

We share many aspects of reproduction with many other organisms. We reproduce sexually, we have chromosomal sex determination (Chapter 7), parental care (Chapter 8) and a diplontic life cycle (Chapter 2). We also practise some forms of courtship, although, uniquely in the living world, this can now happen through the screen of a smartphone. However, in some way (or in many ways, depending on the viewpoint) we are special, or at least we think we are. This impression is not totally unjustified. Beyond any form of human chauvinism (e.g. 'we write books about other species, but no other species writes books about us'), there are two ways in which we are objectively special: one is generic, the other ... special. We are special in the generic sense that every species is special. In this book, we will explore what is common among the ways organisms reproduce, but in the end every species is a separate case, and generalizations can miss some unique feature of reproduction in any given species. The other, special sense in which our species is special is that this is our own species. When we deal with human biology, we are dealing with ourselves, and the way we can study, investigate and understand what we do has no comparison with what we can do on other species. For instance, physicians, psychologists and sociologists use in their studies the answers given by a sample of subjects to a set of questions they are asked. We also assume that important aspects of perception and consciousness are shared among all of us. These aspects are an intimate component of any biological phenomenon that regards ourselves, but their study and understanding cannot be extended to any other living beings.

The relatively simple question of how many sexual conditions there are in a given species can have a straight answer for squirrels (two: male and female), for garden snails (one: hermaphrodite) or for certain populations of the wheel cactus (*Opuntia robusta*) (three: male, female and hermaphrodite). The sexual condition is defined on the basis of the kind of gametes an individual organism can produce (sperm, eggs or both, Chapter 4), and in principle there is no ambiguity in assigning a given individual to one of the three categories, male, female or hermaphrodite. However, as you can imagine, the two simple questions of how many sexual conditions can be recognized in humans and, for a given individual, to which sex does s/he belong, can touch on very delicate issues that are the subject of psychology, anthropology and sociology. In other species, an individual with an anomalous combination of male and female features, or with sex-specific features in an intermediate form, is labelled a gynandromorph or an intersex (Chapter 7). We can study the genetics, morphology, physiology and behaviour of these individuals. For instance, a gynandromorph zebra finch, male on the right side, female on the left side, sang a fully masculine song and courted and copulated with a female (in birds, the right brain hemisphere controls social behaviour). However, we cannot ask the bird how it felt, or how it identified itself.

In most separate-sex species, the sex condition of a given individual can be complicated by the fact that both the establishing of the sex of an individual (sex determination) and the emergence of its sex characters (sex differentiation) are developmental processes, often very complex (Chapter 7). Thus, it is possible that the 'genetic sex' of an individual (e.g. its sex-chromosome complement) is at odds with the titre of sex hormones circulating in the blood ('hormonal sex'), or with the structure of the gonad ('gonadal sex'). Moreover, some sexual traits can be intermediate between male and female, or the individual may show a mosaic of male and female traits, either morphological or behavioural. This is already complex enough; however, in our own species additional questions emerge, involving psychological and sociological factors.

To treat these questions in humans, a common terminological contrivance is to distinguish between sex and gender, where sex includes 'biological' attributes like genitalia, sex-related chromosomes and hormones, while gender includes the individual's psychological and social attributes associated with maleness and femaleness. Gender identity is about how a person identifies, but does not come in binary form, since gender embraces a broad spectrum of possibilities, and a person may identify at any point within this spectrum or even outside of it. It is also recognized that gender identity can change multiple times during a life. Transgender, non-binary, gender-fluid, gender-neutral are only examples of the variety of ways in which a person may define their own situation or viewpoint. Gender orientation describes an individual's physical and/or emotional attraction to other individuals, where heterosexuality, homosexuality and bisexuality are the most common options. Gender is also a social affair. Belonging to a given gender category entails different expectations or behavioural models of, say, masculinity and femininity in different societies, and some people do not identify with some, or any, of the features associated with their gender. Although in most societies there is a basic division between gender attributes assigned to males and females, respectively, some societies have one or more additional gender categories, as some anthropologists and sociologists have described. However, practical as it may seem, the distinction between sex and gender is not universally accepted. Some scholars see a difficulty separating the variables that should discriminate between the physical and the psychological spheres, and tend rather to stress the intimate connection between the two. Gender identity and orientation presumably emerge from a multitude of bodily aspects via (poorly understood) interactions with environment and experience. In any case, the term 'gender' should not be used in non-human species.

Whatever gender may mean, we will not touch on 'gender questions' in this book. This is not because they are not part of our biology, but because they involve mental faculties, such as consciousness, self-perception, intention and will, which, although possibly not exclusive to the human species, are not accessible to investigation in the same way in other species, or through the same analytical tools applicable to our species. It is a question too specific to our own species, and this book is about what is common among the ways all organisms reproduce, with no special focus on human reproduction and sexuality. However, we need to clarify two things, to stress the difference between discussing sex and related matters in our species and in all the others.

First, as a sort of disclaimer, the contents of this book should not be taken as a biological foundation supporting any particular moral, social or political arguments about gender questions, human rights or related medical practices. Observing and studying nature is always informative, but we should be careful not to take nature as a model for humans. Thinking that nature, which in this case equates to the biology of other organisms, can provide a guide for our moral judgement or for taking decisions about our social ordering is a logical error so common that it has a name – the naturalistic fallacy or fallacy of the recourse to nature. This may prompt questions of whether homosexuality, polygamy or incest (but also xenophobia or aggression) are natural, or if they are not instead the (degenerate) product of our culture. In doing so, a double mistake is made, by separating nature and culture, forgetting that culture is an intimate part of human nature, like flight to a sparrow, and by picking up and giving credit to the natural cases that support our (pre)moral judgements, while ignoring those that contradict them. Just to mention the three examples above, homosexuality has been described in some species (e.g. the Chilean flamingo) but not in others; there are species where mating promiscuity is the norm (e.g. the chimpanzee), while in others a pair-bond lasts forever (e.g. swans); there are cases where mother–son mating (e.g. some roundworms) or brother–sister mating (e.g. some mites) is the norm, but this does not occur in all species. But there are also siblings that eat each other (fratricide, many raptors), offspring that devour their mother (matricide, some spiders), females that kill and eat their partner (mariticide, mantises), males in a group that harass a female to mate with her (group rape, some whales), and this list could go on. What should we take from this for ourselves? Nothing, probably, other than an appreciation of some unexpected facets of the diversity of life's phenomena.

Second, the language we adopt follows the scientific use of the terms involved. When we illustrate in later chapters the possibility of 'sex without reproduction' and 'reproduction without sex', one might mistakenly understand that we are talking about contraception and in-vitro fertilization, respectively. But what we actually mean here is that there are sexual processes not associated with reproduction, and modes of reproduction that do not involve sexual processes. When we say 'sexuality' we are referring to sex process, not to the looser behavioural concept of the 'sexual sphere' in our species. And, again, the terms 'abnormal' and 'normal' do not have any connotation as pathological versus non-pathological (as in clinical medicine), and do not take any aesthetic or ethical value (as less-than-ideal versus ideal). The term 'normal' generally refers to the most commonly encountered variant among a set of possibilities, corresponding to the modal type/condition in statistics. This is the way a sentence like 'in separate-sex species, gynandromorphs are abnormal individuals with a mix of male and female traits' should be understood.

That said about ourselves, let's zoom out again, to regain a wider whole-tree-of-life perspective.

The Theoretical Minimum

The great Russian physicist Lev Davidovich Landau (1908–1968) called 'theoretical minimum' everything a student needed to know to work under his tutorship. Landau was a very demanding professor: his theoretical minimum meant just about everything he knew, which of course no one else could possibly know. Thankfully, years later Leonard Susskind and George Hrabovsky used the same term with a different, more accessible meaning. The theoretical minimum that gives their best-selling introduction to physics the title is just what you need to know in order to proceed to the next level. In the same vein, we provide here the minimum of biological knowledge to understand what we will talk about in the following pages. If you have already studied basic biology, you can skip this section.

Living organisms, in a sense that excludes viruses, are made of *cells*: just one, as in bacteria, ciliates and amoebae, which are thus described as unicellular; or many, up to thousands of trillions in a big whale, in organisms which are described as multicellular. In multicellular organisms, many different cell types occur in the body of a single individual, to form specific tissues and organs. For example, no less than 200 types are recognized in the human body.

In terms of structure, cells are of two fundamental types. On the one hand, there is the *prokaryotic cell*, found in bacteria and bacteria-like organisms, which is small (size order 1 μm), with no internal organelles and DNA not separated from the other contents of the cell. On the other hand, there is the *eukaryotic cell*, typical of both unicellular and multicellular eukaryotes; it is larger (size order 10 μm), and with complex content, which includes a *nucleus* (a closed envelope that encloses most of the genetic material), membranes and organelles.

The totality of the genetic material of a cell or of an organism, in the form of DNA molecules, is its *genome*. In eukaryotic cells, the DNA molecules in the nucleus, or *chromosomes*, carry by far the most conspicuous part of the genome, but not its totality. In addition to the nuclear genome there are smaller genomes associated with two kinds of *organelles*: the *mitochondria*, the site of energetic processes known as cellular respiration, which are present in virtually all eukaryotes, and the *plastids*, characteristic of plants, including algae, which contain chlorophyll or other pigments involved in photosynthesis. These organelles are the remnants of bacteria, originally free-living, that have been incorporated a long time ago by a larger cell – within which they survive in a relation of symbiosis, preserving a degree of autonomy in function, replication and transmission across generations, as shown by the persistence of some part of the original bacterial chromosome.

Chromosomes are described as carriers of *genes*, segments of DNA involved in the development of traits. Alternative forms of the same gene (or, equivalently, at the same genetic locus) are called *alleles*. The whole set of gene alleles in a cell (or in the organism) is its *genotype*. The latter term is contrasted to *phenotype*, the whole set of its observable characteristics, morphological especially.

An important feature of the organization of the genetic material is the number of complete sets of chromosomes in the nucleus, which are generally one or two. *Homologous chromosomes* in the different sets have the same genes, although each gene can be present with different alleles. In a human cell, for examples, there are two sets of 23 chromosomes, and thus we are *diploid*; but the unicellular alga *Chlamydomonas* is *haploid* (one set of chromosomes) and the parthenogenetic Indo-Pacific gecko (*Hemidactylus garnotii*) is triploid (three sets). A condition with more than two sets is called *polyploidy*. A cell, or an organism, that presents the same alleles at a given locus of the homologous chromosomes is said to be *homozygous* at that locus; alternatively, it is said to be *heterozygous*.

In prokaryotes, cell division occurs simply by cell fission after the DNA has been duplicated, but in eukaryotes the division of the nucleus, followed by the splitting of the cell, can follow two fundamentally different routes (Figure 1.4). With *mitosis*, chromosomes are first replicated and then equally distributed into two new nuclei, followed by the division of the cell. The two resulting cells are genetically identical and the total number of sets of chromosomes is maintained. In *meiosis*, after chromosome replication, two rounds of chromosome separation follow, which can result in four nuclei, and generally in as many cells, each with half the original number of chromosome sets. In the sexual reproduction of diploid organisms, like us, meiosis gives rise to the haploid egg and sperm from diploid germ cells. Importantly, during meiosis, there is reshuffling of genetic material (*recombination*) so that the chromosomes in the gametes are a mix of the genetic information contained in the two sets of chromosomes, which as a rule were independently inherited by the two parents. Both mitosis and meiosis feature prominently in the life cycle of most

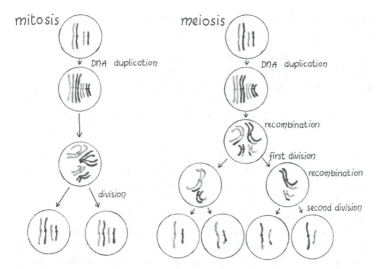

Figure 1.4. Comparison between mitosis and meiosis for a diploid nucleus with four chromosomes. Black and grey colours indicate the distinct parental origin of the two chromosome sets.

eukaryotes (we will have a lot to say about that in this book), but there is another fundamental process that allows the re-establishment of the original chromosome number after meiosis. This is *karyogamy*, the fusion of two nuclei and their chromosomes, as typically follows the fusion of gametes, or *syngamy*. The resulting cell, called a *zygote*, can be the founder cell for the development (through mitosis) of a multicellular organism. As we will see, in most animals, meiosis is the process by which eggs and sperm cells are produced, but in other kinds of organisms the relationships between meiosis and karyogamy are not the same.

In the next chapter we will take a look at the context in which reproduction occurs, the organism's life cycle.

2 Reproduction in the Life Cycle

Monogenerational and Multigenerational Life Cycles

We are all familiar with the changes in an organism during development, followed by its reproduction, which are repeated generation after generation. Biologists describe this development–reproduction sequence as the life cycle: the series of transformations and reproductive events that, from a given stage of life of an organism, leads to the corresponding stage in a subsequent generation. We can describe a biological cycle as going from zygote to zygote, but also from adult to adult, or from embryo to embryo: in a cyclical process, the choice of the 'initial phase' is arbitrary or conventional, as the notorious 'the chicken or the egg' dilemma beautifully illustrates.

In general, however, there are no solid reasons to abandon tradition, and that means starting the description of the life cycle of a multicellular organism with its single-celled stage. By way of example, let's take the life cycle of the fruit fly *Drosophila melanogaster* and move from the fertilized egg, the zygote (Figure 2.1). Within its thin elastic shell, the complex and highly coordinated morphogenetic and gene-regulation processes of embryonic development lead to the formation of a vermiform larva. At 25 °C, embryonic development takes about 12 hours. During the subsequent free-living larval period (about four days, at 25 °C) the insect grows, feeding on rotten fruit, and renews its exoskeleton twice, about 24 and 48 hours after hatching, so that the larval period is divided by the moults into three stages. With a further moult, the third-stage larva transforms into a pupa, an apparently inactive stage that undergoes a process of profound transformation, during which large parts of the larval body are literally demolished, while the adult body, with a completely new organization,

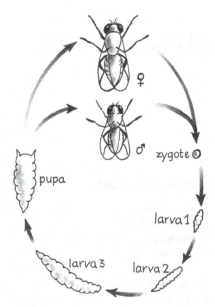

Figure 2.1. Life cycle of the fruit fly *Drosophila melanogaster*, an example of a monogenerational life cycle. Despite the complexity of the developmental process, which includes the metamorphosis of the pupa into the adult, the cycle is closed with one generation and one reproductive event.

builds up. This radical transformation of the individual, called *metamorphosis*, lasts four days, which are spent inside the puparium, a thin envelope formed by the cuticle of the last larval stage, which is not abandoned until metamorphosis is complete. The insect emerges from the puparium as an adult fly. Fruit flies reproduce sexually: males and females mate, and from the fusion of their gametes are produced the zygotes of the next generation. The development of a fly from zygote to adult is complex, because metamorphosis separates two very different segments of the insect's post-embryonic life, the larva and the adult. However, compared to what happens in other organisms, the life cycle of the fruit fly is relatively simple, because the whole story has as its protagonist a single individual that develops and reproduces. This is not the case in a multitude of plants, animals, fungi and microorganisms.

In this sense, ferns, for example, have a more complex life cycle (Figure 2.2). The plant we usually call a fern is only the most conspicuous of the two multicellular phases in the life cycle of these organisms. This phase is called a sporophyte, that is, a diploid plant capable of reproducing by spores. The spores are haploid cells produced by meiosis like the gametes but, unlike these, they are not destined to merge with a complementary cell. Instead, after a phase of dispersion, the spores germinate on the ground, each developing into a tiny haploid multicellular plant called a prothallus. To distinguish it from the sporophyte, this phase is called a gametophyte. The prothalli, which bear both male and female reproductive organs, reproduce sexually, by producing gametes which will fuse to form diploid zygotes, the founding cells of the next cycle's sporophytes. During early development, the sporophyte is retained on

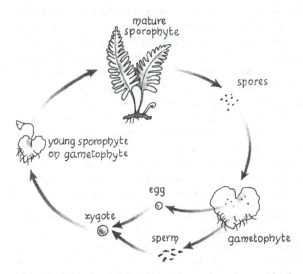

Figure 2.2. Life cycle of the fern *Polypodium vulgare*, an example of a multigenerational life cycle. In this case the life cycle obligatorily traverses two generations separated by two reproductive events, where offspring are generated that are not of the same organizational form as the parents. In the case of the fern, the daughter plants of a sporophyte are not sporophytes themselves, but gametophytes which, in turn, will produce the next sporophytic generation.

the parent gametophyte, which nourishes it until it produces the first leaves and roots and becomes independent. In the life cycle of a fern there are at least two generations (a sporophyte and a gametophyte), which constitute two distinct organizational forms, that is, two kinds of individual of the same species, each with its own development, from unicellular to multicellular, and its own reproduction. One form starts from a zygote to develop into a macroscopic leafy diploid plant, the other form starts from a spore to develop into a tiny haploid prothallus. The two generations are separated by two reproductive events, both sexual: the production of spores by the sporophyte and the fusion of gametes produced by the gametophyte.

The life cycle of the fruit fly is an example of a *monogenerational* life cycle: after one generation, the same sequence of stages of the previous generation is repeated – egg, larva, pupa, adult, all part of the same, single organizational form. The life cycle of the fern is instead a *multigenerational* life cycle: it takes two (in other organisms, even more than two) generations to reach the same developmental stage again. In multigenerational life cycles, the cycle obligatorily traverses more than one organizational form, and thus there are reproductive phases where offspring are generated that are not of the same kind (of the same organizational form) as the parent(s). In the case of ferns, the daughter plants of a sporophyte are not other sporophytes, but gametophytes, which, in turn, will produce the next sporophytic generation. The idea of reproduction as the process through which living beings 'make copies of themselves' holds only up to a point in these cycles with alternation of generations. As we will see in the next sections of this chapter, multigenerational life cycles are far from rare, not only in plants but also in animals and in other large branches of the tree of life.

To understand modes and times of reproduction, we must place them in the context of the other events of the life cycle, like development. However, as biology students are well aware, the study of the diversity of life cycles, fascinating as it might be, is often tricky. In the graphic representations of the life cycles of many groups of animals, plants or protists, one is confronted with the names of stages of development or parts of the reproductive system that do not have an obvious equivalent in other groups, and with a maze of arrows connecting multiple options in development or reproduction, not to mention the simultaneous representation of the macroscopic events at the level of the body of the organism alongside the molecular events at the level of the germinal cells or their genome. But that's life. In the words of a famous misquote from Albert Einstein (actually, a paraphrase of a passage from a 1933 lecture of his), 'everything should be made as simple as possible, but not simpler.' In the same vein, we will now see the most relevant ways to characterize life cycles in the light of reproductive phenomena.

Chromosome Sets Through the Life Cycle

Since the nineteenth-century discoveries of the cellular nature of gametes and the identification of the chromosomes in their nucleus (Box 2.1), the cyclical variation in the nuclear content of the cells of an organism, with the halving of chromosome number at meiosis, and the doubling of their number at fertilization, has been considered a fundamental feature of life cycles. Not surprisingly, for eukaryotes that reproduce sexually, three main types of life cycle are traditionally distinguished, based on the relative predominance of nuclear phases with either haploid or diploid cells: haplontic, diplontic and haplodiplontic.

We find an example of the first type in the *haplontic* life cycle of a unicellular freshwater green alga, *Chlamydomonas reinhardtii*, which for many years now has found a place in the list of prominent 'model organisms', a small selection of species from distant branches of the tree of life that are intensively studied, with the most advanced techniques, to understand a variety of biological phenomena. This alga is almost always found in haploid condition through its life cycle. Its vegetative cells reproduce asexually, typically undergoing 1–3 cycles of binary cell divisions within the parent cell wall, thus producing two, four, or eight daughter cells by each event. However, when the environmental conditions deteriorate (low availability of nitrogen and/or light), these cells move on to sexual reproduction. Since vegetative cells are haploid, it does not take a meiosis to produce gametes; they differentiate into gametes of two alternative mating types (as we will see in Chapter 4, the mating type is somewhat analogous to the sexual condition, for organisms that do not produce distinct male and female gametes). Two gametes of opposite mating type (denoted by the symbols + and –), which are identical in appearance, both with two flagella, can fuse to form a diploid zygote. Zygotes develop over

Box 2.1 Cell Theory and Reproduction

As recorded in any standard history of biology, a really momentous event in the life sciences was the formulation, by the German zoologist Theodor Schwann (1810–1882), of the so-called cell theory (1839), according to which the cell is the common building block of both animal and plant structures. Following this start, the history of major advances towards an understanding of reproduction in terms of cells (prior to the twentieth-century developments in heredity, electron microscopy and, finally, molecular biology) is full of great discoveries that punctuate the decades from the 1840s to the end of the nineteenth century.

First, are gametes also cells, although of a peculiar type? In 1841 Albert von Kölliker (1817–1905) suggested the cellular nature of sperm, and three years later he stated that the mammalian egg must also be considered a cell. The study of sperm structure, however, remained difficult for a long time, given the very small size of male gametes, and it was not until 1865 that Franz Schweigger-Seidel (1834–1871) and Adolph von La Valette-St George (1831–1910) were able to demonstrate that these too, like other cells, are made up of nucleus and cytoplasm.

In 1849 Moritz Wagner (1813–1887) and Rudolf Leuckart (1822–1898) finally succeeded in demonstrating that the intervention of sperm is essential for the egg to be 'activated' and the cleavage process to begin. Fertilization, the union of egg and sperm, was observed for the first time in 1851 by Henry Nelson (1822–1898) in the roundworm *Ascaris lumbricoides*, the large parasitic worm that in the following years became the organism of choice for the study of nuclear division. In 1879 Hermann Fol (1845–1892) demonstrated that only one sperm is needed for the fertilization of the egg. Finally, Leopold Auerbach (1828–1897) observed in 1874 the fusion of the male gametic nucleus with the corresponding female gametic nucleus, in the course of fertilization. In 1883 Édouard Van Beneden (1846–1910), also studying the roundworm, observed how fertilization leads to a doubling of the chromosomal number, since the chromosomes carried by the nucleus of the sperm are added to those carried by the nucleus of the egg.

Thus, there is a problem. In normal cell division (*mitosis*) the chromosomal number is regularly conserved, but this number doubles at fertilization. So why does the number of chromosomes not double with each generation? The solution came from the researches of Theodor Boveri (1862–1915) and August Weismann (1834–1914): animal gametes are produced by a peculiar form of cell division, by which a cell with a certain number of chromosomes produces cells with half that number. This *reductional* cell division was called maiosis by John Bretland Farmer (1865–1944) and John Edmund Sharrock Moore (1870–1947) in 1905; the name was soon changed to *meiosis* by Max Koernicke (1874–1955).

several days into highly resistant dormant spores (zygospores), which can remain viable in the soil for many years and survive freezing and drying out. When environmental conditions improve, and in the presence of light, the zygote undergoes meiosis and typically releases four haploid cells, two for each mating type. These are the unicellular individuals that resume vegetative life. In haplontic life cycles the organism spends almost its entire life in a haploid state, except for the diploid zygote, which immediately undergoes meiosis, restoring the haploid condition. Reproduction is carried out in the haploid phase. Haplontic cycles are not exclusive to unicellular organisms, but are found also in many multicellular fungi (e.g. *Rhizopus*, the black bread mould) and green algae (e.g. *Spirogyra*) (Figure 2.3).

In multicellular organisms, the other two types of life cycle are much more widespread, and in these the relative importance of the phases with haploid or diploid cells is quite different.

In some respects the opposite of a haplontic cycle, in a *diplontic* life cycle the organism is diploid for most of its life cycle, with the exception of the gametes (Figure 2.3). Reproduction occurs in the diploid phase. The life cycle of the fruit fly described above is diplontic, as are the cycles of all animals, including humans. Some brown algae (e.g. *Fucus*), fungi (e.g. *Saccharomyces* yeast), amoebas and many other protists are also diplontic.

The fern cycle described in the previous section is an example of the third, *haplodiplontic*, type of life cycle (Figure 2.3). Here, neither nuclear phase, haploid or diploid (corresponding to the gametophyte and the sporophyte generation, respectively) is fleeting, as a gamete or a zygote can be. Reproduction occurs in both phases, and the sexual processes are distributed between the two reproductive events: recombination at meiosis occurs in the

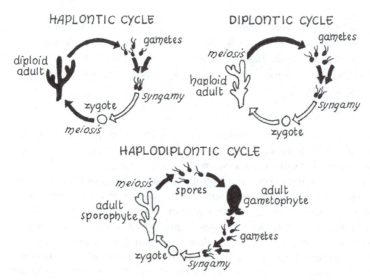

Figure 2.3. Schematic comparison between haplontic, diplontic and haplodiplontic cycles in multicellular organisms. These are depicted as different algal species. The haploid part of the cycle is white, the diploid part is black.

diploid phase, with the production of spores (meiospores), while syngamy involves two gametes produced in the haploid phase. Most plants – in particular, all seed plants, in addition to ferns and mosses – are haplodiplontic, as are many brown algae, some fungi, and some protists.

All multicellular haplodiplonts are plants, and this is reflected in the nomenclature used for the different generations corresponding to the two nuclear phases. The haploid and diploid generations are commonly referred to as the gametophyte and the sporophyte (*phyton* is Greek for plant), respectively, to indicate the haploid generation which produces the gametes and the diploid generation which, by meiosis, produces the spores. We will see the most common haplodiplontic cycles in the next section.

All haplodiplontic cycles include at least two generations, gametophyte and sporophyte. However, one or the other nuclear phase can include multiple asexual generations: in the delicate ferns of the genus *Hymenophyllum*, there are multiple asexual generations both in the haploid phase and in the diploid phase; in the peat moss (*Sphagnum*) this is limited to the gametophyte; in some bamboos (e.g. *Bambusa*) it is limited to the sporophyte.

Note that in diplontic organisms, for example all animals, meiosis (with the production of gametes) precedes syngamy (which re-establishes the diploid condition), whereas in haplontic organisms, for example many green algae, the order of the two sexual processes is the reverse: syngamy (with the production of a zygote) precedes meiosis (which re-establishes the haploid condition). In the haplodiplontic cycles, meiosis and syngamy are carried out by two distinct generations, those of the sporophyte (with the production of spores) and the gametophyte (with the production of zygotes), respectively.

This tripartite life-cycle classification system is well suited to most, but not all, unicellular and multicellular eukaryotes. Notable exceptions are ascomycetes and basidiomycetes, two large groups to which the best-known species of fungi belong. In the life cycle of these fungi, two hyphae with a single haploid nucleus per cell compartment (*monokaryotic* hyphae) can merge, starting what can be viewed as a very long sexual process, but the nuclei they carry remain separate. This produces a first binucleate hypha, which proliferates, giving rise to a mycelium formed by hyphae with two haploid nuclei per cell compartment (*dikaryotic* hyphae). This dikaryotic mycelium proliferates in turn, forming fruiting bodies (mushrooms) that carry the reproductive structures (sporangia) specific for each group. In these specialized structures, generally borne under the cap of the mushroom, the fusion of the two haploid nuclei of the same cell compartment (karyogamy) finally occurs, resulting in a diploid zygotic nucleus. Later, by meiosis of these diploid nuclei, haploid spores are produced that will give rise to the monokaryotic mycelia of the new generation (Figure 2.4). Therefore, in these fungi, between the haploid and diploid phases, there is a third, long-lasting *dikaryotic nuclear phase*, where two separate haploid nuclei are found in the same cell compartment. In most eukaryotes the dikaryotic phase is short and transient, since the fusion of the nuclei immediately follows the union of the cells from which they derive. This is different in ascomycetes and basidiomycetes. When we eat boletus or common table mushrooms, we are actually eating mainly hyphae with two nuclei in each compartment.

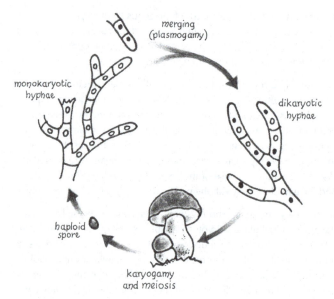

Figure 2.4.   Life cycle of a basidiomycete fungus, an example of a life cycle with a non-transient dikaryotic phase.

Common Haplodiplontic Cycles in Plants

A gametophytic generation alternates with a sporophytic generation in the life cycle of most plants. However, this alternation can all too easily go unnoticed, for different reasons in different groups.

As a first example, let's consider the sea lettuce (*Ulva*). In this common and showy green alga, gametophyte and sporophyte are indistinguishable to the naked eye. The difference, very important but not detectable by morphological inspection, lies in their cell nuclei: the gametophyte is haploid, like the spore from which it derives, while the sporophyte, which takes its form from a zygote, is diploid. This kind of haplodiplontic cycle, with morphologically similar gametophyte and sporophyte, is found in many other green algae, and in some brown algae and fungi.

More often, however, gametophyte and sporophyte of the same species are distinctly different, as in ferns. Here too, the alternation of generations can go unnoticed, but for completely different reasons than in the case of the sea lettuce.

In mosses, for example, the gametophyte is the dominant generation. Haploid gametophytes are the green moss plantlets, on which male and female reproductive organs (*antheridia* and *archegonia*, respectively) generally occur at the tips of different shoots of the same plant. Antheridia produce motile male gametes, archegonia immobile female gametes. Sperm can reach the egg cells by swimming in a film of water that might temporarily cover the plant, and upon fertilization a zygote is formed, which develops into a sporophyte, that is, into a diploid multicellular individual. A moss sporophyte has the shape of a long filament (seta) with a sort of container (urn) at the top. The urn contains cells capable of dividing by meiosis, producing haploid spores. Seta and urn are therefore the diploid multicellular phase in the life cycle of bryophytes. It should be noted, however, that the egg cell, when fertilized, is not physically separated from the gametophyte that produced it, and therefore the resulting sporophyte remains attached to the gametophyte (the mother generation), and often totally depends on this. Consequently, when we observe a moss plantlet complete with filament with a terminal urn, we have before our eyes a composite organism formed of two distinct individuals, one haploid, the other diploid, belonging to two successive generations, the second 'grafted' onto the first.

In spermatophytes, on the contrary, the sporophyte is the dominant generation, with the gametophyte reduced to a few cells and totally dependent on the mother plant, a sporophyte, for its nutrition and development. This is a third reason, specific to seed plants, why the alternation of generations may go unnoticed. In a pine, for example, the sporophyte is the tree, which carries male and female reproductive organs (cones) containing cells (spore mother cells) undergoing meiosis to produce male and female spores, respectively. In male cones, spores develop into tiny male gametophytes (pollen grains) made up of four cells, one of which will function as a male gamete. The spores produced in the female cones develop into female gametophytes that are much larger than the male gametophytes (up to several thousand cells) and never abandon the cone. Pollination occurs before the female gamete is mature, and it may take more than a year between pollination and fertilization. Although two or three eggs can be fertilized within a single female gametophyte, only one zygote develops into an embryo. The complex of the female gametophyte plus the closest tissues of the sporophyte surrounding it is called an *ovule*. Upon fertilization of the egg cell, the ovule will develop into a seed. Most of the life of the male gametophyte and all of the life of the female gametophyte is spent inside the female cone of the parent sporophyte. Thus, seed plants have alternation of generations too: pollen grains and female gametophytes are not the reproductive organs of a mother plant, they are actually its offspring, with a different genotype (derived from meiosis) and a different way of reproduction (through gametes, rather than spores).

In the haplodiplontic cycles the alternation of generations is in register with an alternation of nuclear phases. In the following sections we will see cases in which different organizational forms of the same organism (all diploid) and different types of reproduction are found within a diplontic cycle.

Alternation of Sexual and Asexual Generations

Scientific progress rests on careful observation and well-planned experiment, but also requires insightful and often innovative categorization of the available data. Major conceptual advances are the lasting heritage we got from works such as Darwin's *On the Origin of Species*, or the essay in which Schwann introduced the cell theory. Incomparably less well known than these two major works is a remarkable booklet by the zoologist Japetus Steenstrup (1813–1897), first published in Danish in 1842 but reissued in English in 1845 as *On the alternation of generation or the propagation and development of animals through alternate generations*. Let's rescue it from long oblivion. In that work Steenstrup described in all necessary detail his original observations on some parasitic flatworms (flukes), as well as those of previous authors on the reproduction of some marine animals. Shortly before Steenstrup published his essay, a reproductive link between some kinds of polyps and some kinds of medusae had been described by Sars and Siebold. However, none of these authors or other zoologists before Steenstrup had realized the existence of 'the remarkable, and till now inexplicable natural phenomenon of an animal producing an offspring, which at no time resembles its parent, but which, on the other hand, itself brings forth a progeny, which returns in its form and nature to the parent animal, so that the maternal animal does not meet with its resemblance in its own brood, but in its descendants of the second, third, or fourth degree or generation.' Instead, those facts 'have generally been looked upon as instances of metamorphoses or transformation, the essential objection being overlooked, that a metamorphosis can only imply changes which occur in the same individual; but when from it other individuals originate, something more than a metamorphosis is concerned.' Steenstrup introduced the term 'metagenesis' to describe the alternation of generations between polyp and medusa, and it is still used mostly to describe this particular case – but it can also be used with a wider meaning.

Most jellyfish, or medusae, have separate sexes, and sexual reproduction is generally amphigonic, with internal or external fertilization, depending on the species. The zygote develops into a planktonic larva which, after a wandering pelagic phase, attaches itself to a substrate where it will transform (metamorphose) into a polyp. This is a different kind of organism than the medusa, sessile (i.e. permanently attached to a substrate) rather than planktonic, although from the point of view of body architecture it could be seen as an upside-down medusa, with tentacles protruding opposite to where it is attached and the anchor base corresponding to the top of the medusa bell. The polyp, offspring of a male and a female medusa, does not reproduce sexually. After a growth phase, it generates a certain number of medusae asexually. In turn, the latter will grow and, upon reaching maturity, will reproduce sexually, producing gametes which through fertilization give rise to the zygotes of the next cycle (Figure 2.5). In this case, the alternation of generations is not accompanied by alternation of nuclear phases: both the polyp and the medusa are diploid. This is an example of a *metagenetic cycle*, a multigenerational cycle in which strictly asexual generations alternate with sexual generations, represented by distinct organizational forms. Metagenetic cycles are distinguished from other multigenerational cycles by the fact that there is at least one obligate asexual generation, morphologically and physiologically distinguishable from the sexual generations with which it alternates.

Many representatives of the cnidarians have a metagenetic life cycle with alternating generations of polyp and medusa. In some cases, the medusa is more conspicuous than the polyp; in others the polyp, which can form

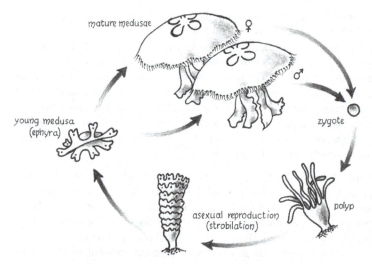

Figure 2.5. Life cycle of the cnidarian *Aurelia aurita*, an example of a metagenetic life cycle. The life cycle obligatorily traverses two generations, one reproducing sexually (the medusa), the other reproducing asexually by strobilation (the polyp).

colonies, is more conspicuous than the medusa, which mostly functions as a short-lived dispersal phase. In other cycles, either the medusa or the polyp is missing. In the first case, the polyp generates other polyps sexually. To enrich the variety of cycles, asexual reproduction is widespread: depending on the species, it involves the production of polyps from polyps, medusae from medusae, or (as generally required to close the cycle) medusae from polyps.

Metagenetic cycles offer a splendid opportunity to show how the way we classify things can affect our perception of the life cycles of organisms. Let's for a moment turn our attention to twins. When more than one embryo develops from a single zygote, we say that there is polyembryony (Chapter 3). At some point in embryonic development, generally very early, the embryo divides into two (or more) genetically identical embryos, each with the prospect of developing into an independent adult. This is what happens with identical human twins, an event not rare, but certainly not characteristic of our species.

However, there are species in which polyembryony is obligate, that is, each fertilization event ends with multiple genetically identical offspring. In this case, the splitting of the embryo can be considered as a form of asexual reproduction that occurs at a very early stage of development, practically producing an alternation of sexual and asexual generations. Consequently, we should list some species of flatworms, wasps and armadillos among the animals with a metagenetic cycle.

Alternation of Biparental and Uniparental Sexual Reproduction

In some eukaryotes, a generation that reproduces biparentally through fertilized eggs (i.e. by amphigony) regularly alternates with generations that reproduce uniparentally through unfertilized eggs, that is, by parthenogenesis (Chapter 6). In technical terms, cycles of this type are called *heterogonic*. The best-known examples of these multigenerational cycles are those of water fleas (freshwater cladoceran crustaceans of the genus *Daphnia*) and plant lice (aphids), but they are also found in some species of parasitic roundworms and in most species of a group of tiny aquatic invertebrates known as rotifers. The transition from parthenogenesis to amphigonic reproduction depends on the interpretation of specific signals from the environment, such as the seasonal reduction in day length or an increase in population density. However, there are significant differences in how the environmental signal is received and subsequently interpreted by the organism, to determine its physiological response.

In water fleas, both males and females are diploid. While females are regularly present in the lakes and ponds in which they live, males appear only at certain times of the year, different according to species and latitude. In this group, reproduction by parthenogenesis alternates cyclically with amphigony, when males are present. Parthenogenetic females produce subitaneous (i.e. immediately developing) diploid eggs, which develop into adult females within 3–4 days. However, in response to specific environmental signals, such as the shortening of the day length or an increase in population density, parthenogenetic females switch from the exclusive production of females to the generation of males and females. The environmental signal that elicits the production of males is received by the parthenogenetic mother, who translates it into a hormonal signal which has an effect on the maturation of the egg; in turn, this will affect the prospective sex of the offspring. This is an example of maternal sex determination, where the sex of an individual does not depend on either its genes or its personal perception of the environment, but rather on the genes or the physiological condition of the mother. When males are present, the same parthenogenetic females can also produce haploid eggs that, if fertilized as a result of mating, become resting (i.e. quiescent) eggs. These, protected singly or in pairs in special egg-cases produced by the mother and/or by her old exoskeleton, released at the moult during which she lays the eggs, are able to face adverse environmental conditions. When suitable environmental conditions are encountered, possibly in the following spring, only females will develop from these eggs. In cladocerans, the same female can produce both types of eggs and, after amphigonic reproduction, resume reproducing by parthenogenesis. Some species perform one heterogonic cycle per year, others go through several cycles per year.

Alternation between amphigonic and parthenogenetic reproduction is also found in numerous species of insects, belonging to different groups (e.g. a number of species of the gall midge family); however, it is among the aphids that a particular diversity of heterogonic cycles has evolved.

The aphid life cycle generally lasts one year, but in some species it can extend over several years; in other cases, the amphigonic generation can become sporadic or even disappear, as in some species that live in warm climates. The cycles of aphids are further complicated by the alternation between winged dispersive forms and sedentary wingless forms, and by the possibility of changing host plant, or the part of the plant they feed on, one or more times. In the general case, the following sequence is repeated in an annual cycle: parthenogenetic females generate daughters, also parthenogenetic and generally wingless. At least one, and most often several generations of these females will follow. At a certain point a generation of females appears which by parthenogenesis give rise to an amphigonic generation, with males and females. This amphigonic generation is generally winged. Their fertilized eggs pass through the winter season at rest. A specific mechanism in the maturation of sperm ensures that these fertilized eggs all develop into the parthenogenetic females that will begin a new cycle.

In most aphids the parthenogenetic generations are viviparous (i.e. bring forth live young) and only the amphigonic female is oviparous (i.e. lays eggs), but in some species all generations are oviparous. In many aphid species, the main environmental signals that induce the production of the amphigonic generation are short day length and low temperature. However, in some aphids, amphigonic reproduction occurs in response to physiological changes in host plants, for example in response to the cessation of growth of host-plant shoots. The environmental signal that causes the transition from parthenogenetic reproduction is received and processed by the female who generates the amphigonic offspring.

Cycles with Reproductive Options

The diversity of multigenerational life cycles does not end here. In some nematode worms there are cycles with separate-sex generations and hermaphrodite generations. Many marine invertebrates and green algae have cycles with alternating solitary and colonial generations, and in plasmodial slime moulds unicellular and multicellular generations alternate. Some tropical butterflies have two generations per year, morphologically distinguishable based on the colour pattern of the wings and specifically adapted to the wet or dry season respectively (*seasonal polyphenism*). As is typical of alternations between two organismal forms, butterflies of a generation (say, dry-season morphs) closely resemble their grandparents and grandchildren, but are not as similar to their parents and children (wet-season morphs).

In summary, multigenerational life cycles include multiple organizational forms that can exhibit different genetic make-up (e.g. haploid versus diploid in mosses), different morphology (e.g. winged versus wingless in aphids), different living environment (e.g. different host in parasitic flatworms), a different mode of reproduction (e.g. sexual versus asexual in pelagic tunicates), and/or different type of development (e.g. with or without metamorphosis in cnidarians). There is no doubt that in many organisms the path through which the life cycle closes onto itself can be very tortuous.

The life cycle of some species, however, is even more complex, because in certain situations there is more than one option available, to continue development or to reproduce. The 'choice' generally depends on the physiological state of the organism at a given time, or on the occurrence of specific environmental conditions. Alternative life-cycle pathways may first diverge and then converge again, in a subsequent developmental stage of the same generation or in a subsequent generation, eventually producing completely equivalent organisms. What happens in these cycles is an instance of the more general phenomenon of phenotypic plasticity, where individuals with the same genetic make-up can express different phenotypes as a response to the different environmental conditions in which they develop. Many examples of life cycles with developmental and/or reproductive options are found among parasites, roundworms and flukes especially. Given the subject of this book, let's focus on reproductive options.

Take the case of the cellular slime mould *Dictyostelium discoideum*, also known as the social amoeba. In the asexual cycle of this species, generations of solitary, haploid amoeboid cells follow each other, feeding on soil bacteria and reproducing by binary fission. When food is scarce, many cells aggregate, mutually exchanging chemical signals, entering the so-called aggregation phase: they form a mass, at first approximately discoid, which turns into a migrating 'slug' (which is just what it looks like, except for its tiny size). Its wandering soon stops, and the slimy slug turns into a pedunculated fruiting body, a sort of tiny mushroom. The cells at the top of this become resistant spores which are released and dispersed into the environment. At a later time, under favourable conditions, the spore wall breaks, freeing an amoeboid cell that returns to the trophic phase and to asexual reproduction. In this asexual cycle, reproduction occurs during the solitary phase, since there is no cell proliferation during the aggregate phase. In simple terms, a few cells become spores and may have descendants, while all the others die. However, free-standing individual amoebas can also switch to a sexual cycle. In this case, two amoebas merge and form a zygote that grows by attracting and engulfing others of the same species (a form of cannibalism). This giant zygote develops a protective wall and becomes resistant. It undergoes meiosis, followed by several mitotic cycles, whereby several haploid amoeboid cells derive, ready to resume the trophic phase.

Generally, reproductive options occur when a given reproductive modality is qualified as facultative, optional or cyclic, rather than obligate or constitutive. Parthenogenesis is facultative in many molluscs, annelids and arthropods, and also in some vertebrates, including the Komodo dragon (*Varanus komodoensis*). In the latter species, facultative parthenogenesis has been demonstrated in captive specimens. Owing to the cost of breeding such huge reptiles, zoos often keep only a single animal. In several instances, females kept isolated from any other member of the species laid eggs from which healthy offspring emerged, after a very long incubation (more than seven months).

Self-fertilization (Chapter 6) is facultative in various hermaphrodite animals, including some pulmonate gastropods (e.g. terrestrial snails of the genus *Rumina*), while self-pollination (which is distinct from self-fertilization) is facultative in a number of plants, including various members of the legume, orchid and aster families. Likewise, asexual reproduction is facultative in many organisms that usually reproduce sexually. In some animals, reproduction can be carried out both by the adult and by a juvenile stage (paedogenesis, Chapter 6), as in the larvae or pupae of some insects.

The choice between alternative reproductive options is generally influenced by specific environmental factors, which differ from species to species. For example, animals that exploit ephemeral food sources may adopt different forms of reproduction in response to the different availability of nutritional resources. When food is abundant, many species of gall midges reproduce by parthenogenesis when they are still larvae (paedogenesis), but when food is scarce they reproduce as adults instead. In one of these midges, *Heteropeza pygmaea*, when the availability of food is not optimal, the larvae develop into adults, males and females, which fly in search of new food sources. There they mate, and the females lay their eggs, only some of which have been fertilized. From the unfertilized eggs, with the benefit of the abundant food provided by the newly colonized mushroom, a new generation develops, all females, which reproduce by parthenogenesis when they are still larvae, without laying eggs. The daughter larvae develop in the body cavity of the mother, where they complete embryonic development before the mother dies. In turn, these larvae can generate other female larvae that continue the paedogenetic cycle, or, if food resources become scarce, generate larvae that develop into adult males and females.

Summing up, in this chapter we have surveyed the diversity of life cycles (Table 2.1), including cycles that can follow alternative paths. We are now ready to examine the different modes of reproduction in more detail.

Life cycle
- monogenerational
- multigenerational (with alternation of generations)
  - alternation of haploid and diploid generations (haplodiplontic)
  - alternation of sexual and asexual generations (metagenetic)
  - alternation of amphigonic and parthenogenetic generations (heterogonic)
  - alternation of solitary and colonial generations
  - alternation of separate-sex and hermaphrodite generations
  - alternation of unicellular and multicellular generations
  - alternation of different morph generations (seasonal polyphenism)

Table 2.1.   Summary of the main types of life cycles.

3 Reproduction Without Sex

Clones and Asexual Reproduction

On February 1997 the birth was announced of a sheep named Dolly, the first mammal to be cloned from an adult cell of a mother individual. The event attracted enormous media attention. Dolly, born on 5 July 1996, actually had three 'mothers': one provided the egg (whose nucleus was removed), another the nucleus with the DNA picked out from a somatic cell (i.e. a cell of the body not specialized for reproduction), while the third mother carried the cloned embryo in her womb until parturition.

The term *clone* is used in biology to say that an individual organism is genetically identical to another individual. The term can also indicate a set of genetically identical organisms, usually implying derivation without genetic modification from a single ancestor. Dolly was a clone of the second of her three mothers, the one who provided the nucleus.

As outlined in Chapter 1, asexual reproduction is a mode of reproduction from a single parent that does not involve sexual processes or the production of gametes, not even in derived or residual form. As a first approximation, asexual reproduction generates individuals genetically identical to each other and identical to the parent, thus forming clones. For this reason, asexual reproduction is sometimes called clonal reproduction. However, the equation asexual = clonal is not exact, since asexual reproduction may not produce perfectly clonal descendants, while sexual reproduction may have a clonal outcome. In this section we will see how genetic variation can arise through asexual generations, while clonal sexual reproduction is covered in Chapter 6.

The main cause of the non-perfect genetic identity among the members of a clone is genetic mutation. Mutations may occur during reproductive processes, or arise in other phases of the life of an individual. A first source of genetic mutations is in the process of DNA replication. DNA replication is an extremely precise process, but not a perfect one. The frequency of copy error is very low (in the order of one-billionth per base pair, the fundamental unit of DNA sequence, per replication), but not zero. In the reproduction of unicellular organisms by binary fission, mutations can make the daughter cells genetically different from one another and (at least one of them) from the parent cell. In the same way, modified cell lines may arise during cell proliferation along the development of a multicellular organism, a condition called *genetic mosaicism*. If a mosaic organism reproduces asexually from cells belonging to one of these modified cell lines, the offspring's genotype will not be perfectly identical to the (original) parental one.

But mutations can result also from other causes. During the proliferation of a clone of unicellular organisms, or during the lifetime of a multicellular organism, the genome of some cells may mutate due to chemical or physical factors of various kinds, the most effective of which are known as mutagens. Mutagens are among the most dangerous environmental pollutants and can represent a serious threat to human health, an example being some polycyclic aromatic hydrocarbons (PAHs) that can cause mutations leading to cancer.

Similar to mutations, some mechanisms of genetic recombination, such as crossing over in eukaryotes, which can occur even during normal cell division by mitosis, can undermine the perfect genetic identity otherwise shared by the members of a clone, or transform a multicellular individual into a genetic mosaic. Another source of genetic variation can be found in mechanisms of multiplication and distribution of hereditary material other than the main chromosome of prokaryotic cells or the nuclear genome of eukaryotic cells. The distribution of these DNA molecules among the daughter cells during cell division is not rigidly controlled, so that the descendant cells may receive unequal genetic material. This happens, for instance, following the random segregation of small DNA molecules (plasmids) at cell division of prokaryotic cells, or the segregation of cytoplasmic organelles (mitochondria and/or plastids), which have their own genomes, in eukaryotic cells.

In summary, although in principle asexual reproduction could be regarded as equivalent to clonal reproduction, actually, this is true only to a limited extent. A large number of processes and events, more or less strictly related to reproduction, make it possible that a certain amount of genetic variation is generated within a clone. In practice, 100% pure clones are very improbable, and when we refer to the genetic identity of the members of a clone, it is implied that we are neglecting variation due to mutations of recent origin. For instance, it is estimated that in a middle-aged human being, with a number of somatic cells in the order of 10^{13}, the 6×10^9 base-pair nuclear genomes of his/her cells can collectively present a number of point mutations in the order of 10^{16} (i.e. 1,000 per cell, on average), most of which, for purely statistical reasons, are in regions of the genome that do not code for proteins (non-coding DNA).

Within-individual genetic heterogeneity, usually considered to be insignificant, may become conspicuous in organisms that are very old or very large, where many rounds of cell division may have occurred. Numerous mutations may have accumulated along the branches of an old willow tree, sitting on the bank of a river. Two young branches, broken away from it and floating some distance downstream, have a good chance of taking root on the river bank, and developing into two young trees genetically different from each other. Furthermore, because of the modular organization of these plants, each terminal branch, with its flowers, can be considered an independent site of sexual reproduction, and thus the genetic make-up of the cells that form pollen and ovules on one branch can be significantly different from the genetic make-up of similar cells on another branch of the same tree. In the case of oaks, which can live longer than willows, two reproductive cells of the same old tree may belong to cell lineages that separated centuries ago. Such distances in terms of time and cell ancestry are comparable to those observable in an entire population of smaller organisms with shorter life, like many insects.

Having discussed the genetics of clones, let's now move on to see how various organisms reproduce asexually.

Asexual Reproduction in Unicellular Organisms

In both prokaryotes and unicellular eukaryotes, asexual reproduction coincides with cell division. The process of cell division, however, differs substantially between the two main groups of living beings, because of the very different organization of the cell, in particular the presence of a nucleus, sometimes even more than one, in eukaryotes.

In prokaryotes, reproduction is always disjoint from sex, and may occur by fission or by sporulation. In *fission*, either binary or multiple, the dividing cell generates metabolically active daughter cells (vegetative cells), whereas in *sporulation* the division gives rise to one or more quiescent cells (spores) capable of resisting adverse environmental conditions. Bacterial spores exhibit exceptional resistance to extreme heat, desiccation, radiation and toxic chemicals. This is why spore-forming pathogenic bacteria, such as *Clostridium difficile* and *Bacillus anthracis*, are highly resistant to antibacterial treatments and are difficult to eradicate. Quiescent bacterial spores may be able to germinate for up to thousands, or even millions of years. A bacterial spore was revived from the gut content of extinct bees preserved in amber for between 25 and 40 million years, and a group of researchers claims to have revived bacteria from salt crystals 250 million years old. These bacteria would be the oldest known living beings. Fission and sporulation are often reproductive options for the same bacterial species, which can switch from one mode of reproduction to the other in response to environmental conditions. When these are favourable, the bacterium undergoes fission; when conditions become adverse, it turns to sporulation.

In unicellular eukaryotes, cell division consists of a number of processes, among which are the division of the nucleus by mitosis and the repartition of cell organelles among the daughter cells. In eukaryotes, as in prokaryotes, we can find variation in the number of the products of division. *Binary fission* is the most common reproductive mechanism among unicellular eukaryotes. At a certain stage of the cell cycle, for example when the cell has reached a certain size, mitosis occurs and the cell divides. Cell organelles, mitochondria and plastids, and cell structures such as flagella or cilia, can be replicated before or after the separation of the daughter cells, paralleling an option we will see in asexually reproducing animals (paratomy versus architomy; see below). There may be an interval between nuclear division and cell division, however; in this case, the cell remains binucleate for a while.

If nuclear division is repeated several times before the cell divides, there is no cell multiplication and therefore no reproduction of a unicellular organism; instead, a multinucleate entity develops. However, the multinucleate condition is often just a prelude to a multiple division into a large number of daughter cells, each of which normally inherits only one of the nuclei deriving from the previous nuclear divisions. This is *multiple fission*, which is frequent among the unicellular eukaryotes, where it is also called *schizogony*. Malaria is an infectious disease that affects humans and other animals, which is caused by a protist of the genus *Plasmodium*, spread through the bites of infected *Anopheles* mosquitoes. The classic symptom of malaria is a cyclical attack of fever, occurring at intervals of two days (tertian fever, e.g. in *Plasmodium vivax* infection), or three to four days (quartan fever, e.g. in *Plasmodium malariae* infection). These feverish accesses coincide with the moment the infected red blood cells of the host break apart as a consequence of the synchronous massive schizogony of the parasite.

Other variants in the patterns of asexual reproduction in unicellular eukaryotes, which we will encounter again in multicellular organisms, depend on the relative size of the products of a division event. In asexual reproduction by *symmetric cell division*, such as in binary fission, the parent's body is divided between two approximately identical descendants. In symmetric binary fission, the two cells that are thus obtained are described as sisters, both descendants of an individual which, by dividing, has ceased to exist. By contrast, in asexual reproduction by *asymmetric cell division*, such as budding, the parent cell persists as a distinct individual across the reproductive act, while a minor portion of its body becomes its offspring. The larger cell is called the mother cell, while the smaller cell that detaches from it is called the daughter cell. The different terms used to describe the products of asexual reproduction through symmetric and asymmetric cell division are justified, in certain cases at least, by their different behaviour with respect to senescence (Chapter 1). In the protist *Euglena*, which divides symmetrically, the two daughter cells, like two twin sisters, have the same life expectancy. However, a mother cell of the common bread yeast *Saccharomyces cerevisiae* buds off daughter cells with a longer life expectancy than her own current value, exactly as it should be for a mother's offspring. However, a morphologically symmetric division can hide a very asymmetric division at the molecular level, with unequal partition of damaged cell constituents (a consequence of ageing) among the daughter cells, so that one of the two sisters is actually 'chemically younger' that the other. In unicellular organisms (but also in multicellular ones), budding, either binary or multiple, is sometimes associated with the formation of colonies, as in some ciliates, where the buds do not detach from the parent body but remain connected, to build a multicellular structure that can take different shapes, from frond-like to feather-like.

Unicellular Propagules in Multicellular Organisms

In multicellular organisms, asexual reproduction occurs when a new individual originates from a piece – either unicellular or multicellular – of the body of the parent, which takes the name of *propagule*.

Unicellular propagules are generally called *spores*. This term is used with many, more or less overlapping, meanings in the biology of unicellular and multicellular prokaryotes and eukaryotes. What is common to most of these spores is the fact that they are reproductive cells which can develop into a new organism without merging with another cell, as gametes do. Moreover, a spore is often a robust resting stage, often corresponding to a dispersal phase of the life cycle.

The spores through which eukaryotes reproduce asexually are produced by mitosis; these spores are either haploid or diploid, depending on the chromosomal asset of the individual that produces them. These should not be confused with the haploid spores produced by meiosis within life cycles of plants and fungi, which are part of the processes of sexual (rather than asexual) reproduction of those organisms (Chapter 2).

Spores can differ greatly in terms of appearance, size and behaviour. Some, such as the zoospores of brown algae, are equipped with flagella and are thus able to move in a liquid medium, while others are immobile, such as the asexual spores (or conidia) of the ascomycetes, produced on specialized hyphae. Some algae, for example *Vaucheria*, produce both mobile and immobile spores. The spores of certain fungi are so small and light that winds can carry them away to a great distance. Fungal spores have been found in air samples collected at an altitude of 100 kilometres. Some fungi, however, themselves contribute to the dispersal of their spores. This is the case in *Pilobolus*, a fungus that commonly grows on the dung of herbivores. Its spore-bearing structure, or sporangiophore, is a delicate stalk rising above the dung that supports a vesicle with a sporangium on the top. The sporangiophore grows oriented towards a light source. When fluid pressure within the vesicle reaches about 7 atmospheres, the sporangium is launched up to a distance of 3 metres.

Asexual spores are found in many groups of algae and fungi, but not in animals and land plants. In these groups, clonal reproduction starting from a single parental cell is almost exclusively practised through some forms of partheno-genesis, as we will see in Chapter 6. A possible exception is offered by two small groups of poorly known parasites of marine invertebrates, which are among the animals with the simplest body organization. Dicyemids and orthonectids have only about 40 and a few hundred cells when adult, respectively, but they have nonetheless complex life cycles with alternation of asexual and sexual gener-ations (metagenetic cycle, Chapter 2). The propagules for asexual reproduction (agametes) are unicellular and develop inside the parental body.

Multicellular Propagules in Plants and Fungi

In many algae and fungi, but also in mosses, asexual reproduction is frequently accidental, by fragmentation. A small part detached from the frond of an alga, the stem of a moss, or the mycelium of a fungus, becomes independent and may grow into a new individual.

Lichens are symbiotic associations between a green alga (less frequently, a cyanobacterium) and a fungus (usually, an ascomycete). These symbiotic aggregates reproduce both sexually and asexually, and both partners, the alga and the fungus, are involved. Asexual reproduction of lichens is relatively simple, because it goes through the production of body fragments in each of which both the fungus and the alga are present: therefore, these propagules can grow into a larger lichen body without the need for external intervention. More complex is a lichen's sexual reproduction, because this begins with the dissociation of the symbiosis between fungus and alga: the former undergoes the path of fungal sexual reproduction until new individuals are formed, each of which will eventually develop into a lichen if it is reached by spores of a compatible algal species.

In seed plants, depending on the species, a new individual sporophyte can take hold from an incidentally removed small twig, from a piece of root, or even from a leaf. In one of his short stories, the twentieth-century Uruguayan writer Horacio Quiroga, when depicting the exuberant vital power of nature in the tropics, beautifully describes the unstoppable tendency to take root and compete for resources of the bamboo sticks that were supposed to support climbing plant crops (beans). In agricultural practice, propagation by cuttings, where a plant fragment is cut and planted, exploits these regenerative capacities in many plants of economic interest.

In seed plants, vegetative reproduction may take forms more specialized than traumatic fragmentation. This may imply detachment of a part of the individual at a specialized body location (abscission zone) or localized decay of tissues separating the surviving parts. Many of these forms of asexual reproduction can be observed in a domestic garden (Figure 3.1). Strawberries normally propagate by *stolons* (or runners). These are specialized, horizontally developing stems in which long and thin internodes (stem intervals between two nodes) alternate with very short internodes capable of sprouting extra (adventitious) roots and therefore giving rise to new plants that will become independent with the death of the stretch of stolon connecting them to the mother plant. *Rhizomes* are similar, but usually subterranean and stouter. Many ferns, many aquatic plants (e.g. water lilies) and many reeds have rhizomes. Lily of the valley and irises also have rhizomes. Alliums, daffodils, hyacinths, grape hyacinths (*Muscari*) and tulips have *bulbs*, short stems with a vertical axis, more or less completely buried and with abundant reserve material located in the foliar bases or in scale-shaped leaves. The buds that differentiate at the axils of these leaves develop into second-generation bulbs, which survive the rotting of the parent bulb and will eventually give rise to new plants. Anemones, caladiums and potatoes have *tubers*, short, enlarged subterranean stems or enlarged roots with adventitious buds (i.e. buds that do not develop in the axil of a leaf, as usual). The tuber provides a way for vegetative multiplication when the long thin connection that initially unites it with the mother plant dries out. Bulbils are small propagules that can develop at different places, for example in the leaf axils (*Lilium*). In

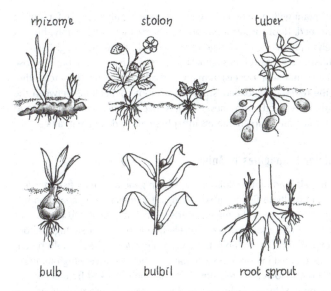

Figure 3.1. The most common types of plant multicellular propagules.

Allium, bulbils often replace flowers inside the reproductive structure that would otherwise be an inflorescence. Bulbils are common among the Liliaceae and Amaryllidaceae, and also in ferns.

These phenomena of vegetative propagation are so widespread among plants that, as we have seen in Chapter 1, botanists have found it useful to introduce two terms to indicate two different kinds of 'plant individuals'. A *genet* is a set of genetically identical entities (which can be considered either individuals or modular parts of an individual) derived by clonal multiplication from a single genetically unique individual. All the apple trees of the 'Red Delicious' variety, which are derived from cuttings of a single tree that lived in Iowa (USA) in the late nineteenth century, are part of a single genet. In contrast, a *ramet* is an anatomically and physiologically bounded biological entity, independent of its genetic constitution. As such, it may well be a member of a genet (each 'Red Delicious' apple tree is a ramet), or an individual plant with genetic uniqueness, as is generally the case for a plant born from seed. Root sprouts are produced by many plants, mainly, but not exclusively, in response to damage to the roots. In some trees (e.g. *Populus*, *Liquidambar*), very large clones (genets) can originate from root buds, whose 'individuals' (incipient ramets) remain partially connected to each other through the roots. In herbaceous plants, the root connections between the individuals developed from these sprouts decay much more easily and quickly, turning the incipient ramets into physically independent ramets.

Multicellular Propagules in Animals

Among animals, and multicellular organisms in general, binary fission is less frequent than multiple fission, if we disregard incidental fragmentation followed by regeneration of the missing parts in each fragment. Observation of regeneration in planarians cut transversely into two is a classic high-school science experiment. In multiple fission (or *schizotomy*), the whole parent individual is divided into a number of parts, sometimes a very high number. Fragmentation, followed by regeneration, may be provoked by mechanical trauma, or induced by poor physiological conditions, for example starvation. The marine ribbon worm *Lineus sanguineus* possesses some of the highest regenerative capabilities known among animals. Individuals of this species can be repeatedly cut into pieces until the resulting worm fragments still able to regenerate are just 1/200,000 of the volume of the original individual. A complete animal can regenerate not only from a thin transverse slice of the body, but even from just one quadrant of a thin slice.

Multiple fission can also be controlled by the organism itself. Two main modes can be distinguished, one in which division follows development, while in the other development follows division (Figure 3.2).

In *paratomy*, a new complete individual is recognizable before its detachment from the parent. In the asexual reproduction of some syllid polychaetes, a single 'individual' formed by a chain of 'individuals' at different stages of maturation can be observed, with the rearmost individual being the most advanced and ready for abscission. Within the polychaetes, paratomy is known in several groups. Asexual reproduction by paratomy is common in

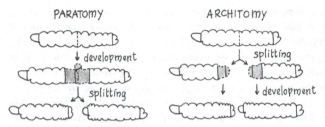

Figure 3.2. Different modes of asexual reproduction occurring in polychaetes. With paratomy, the new individuals complete development before splitting. With architomy, splitting precedes the development of the missing part in each fragment.

catenulid flatworms, usually leading to a chain of developing individuals (hence the name, from Latin *catenula*, small chain).

In *architomy*, instead, the simple fission or fragmentation of the body happens before the whole organization of a new complete individual is recognizable in each fragment. Architomy is widespread among sponges, ribbon worms, flatworms (in particular, freshwater planarians), but also in some families of oligochaetes and in the starfishes. Animals with high regenerative capacities usually resort to reproduction by fission, and those that reproduce by fission also have high regenerative capacity, but this is not a universal rule.

There are also intermediate situations between paratomy and architomy. These include *strobilation*, a term that mainly applies to the subdivision of the polyp of scyphozoans (cnidarians) into a stack of tiny medusae (ephyrae; Figure 2.5) that progressively detach, but also to the way in which the body of a typical tapeworm (e.g. the pork tapeworm, which has humans as its definitive host; Chapter 6) is articulated in segments known as proglottids. Among the scyphozoans there are species in which only one little medusa (ephyra) detaches from the parental polyp (scyphistome), although the number is usually higher and variable between individuals of the same species, being larger for larger scyphistomes.

In metazoans reproducing by *budding*, a portion of the parent's body grows through cellular proliferation and/or reorganization of already differentiated tissues to become a new organism that eventually detaches from the parent. We have already referred to analogous (and homonymous) mechanisms of asymmetric division in prokaryotes and unicellular eukaryotes. A bud can be formed from somatic tissues with cells that have regained multipotency, in other words the capacity to differentiate into a diverse array of cell types. However, this is not an absolute necessity, as exemplified by the case of *Hydra*, where the buds are formed by cells belonging to already differentiated primary cell layers (endoderm and ectoderm). Budding is frequent in sponges, cnidarians (especially among the polyps of the hydrozoans and cubozoans), entoprocts, bryozoans and phoronids.

Similar to many plants, some animals reproduce by *stolons*, elongated off-shoots that remain connected to the parent organism, sometimes permanently. Groups of cells differentiate along the stolon, from which new individuals originate; the latter will separate by mechanical action or by degeneration of the interposed tissue, or remain united together to form a colonial group. Stolons are frequent in sessile marine invertebrates, for example among the cnidarians (colonial polyps), entoprocts and bryozoans, but they are found also among pelagic polychaetes and tunicates.

A remarkable example of stolon organization is found in two polychaete species, *Syllis ramosa* and *Ramisyllis multicaudata*. Unique among the animals with primitively bilateral symmetry (the bilaterians, to which most animals belong), they have an extensively branched body that mirrors the intricate topology of the canals of the sponges within which these worms live. They have only one mouth, but a branched gut with many anal openings. Each branch of these worms is a stolon that instead of breaking away remains attached to its stem and produces more branches.

Mechanisms of asexual reproduction are especially diverse among the colonial tunicates, including budding from a number of internal organs and body regions, production of stolons bearing either many lateral or one terminal bud. Most singular is the mechanism known from the colonial ascidian genus *Botryllus*, where blood cells can aggregate in the vascular network to form one of the so-called vascular buds from which whole-body regeneration may start.

Polyembryony and Larval Amplification

In humans, monozygotic (identical) twins naturally occur approximately once in 300 births, whereas monozygotic triplets or quadruplets are much rarer. Depending on the time the early embryo splits, before or after implanting in the womb, the two fetuses experience different degrees of commonality. They may have independent placentas and amniotic sacs, or share a placenta while having independent amniotic sacs, or share both placenta and amniotic sac. Twin births are relatively rare in humans, although more frequent with in-vitro fertilization. Similarly, twins occasionally occur in almost all major animal groups, including domestic cattle, pigs and chickens. However, in some organisms, the derivation of more than one embryo from a single zygote, a process that takes the name of *polyembryony*, is recurrent, or even the rule.

Polyembryony is common, or even obligate, in representatives of cnidarians, flatworms, bryozoans and insects, but also in some mammals. In armadillos of the genus *Dasypus* polyembryony is obligate, and the number of embryos deriving from a single zygote varies from 2 to 12 according to species. Nine-banded armadillos (*Dasypus novemcinctus*), which always produce litters of identical quadruplets, are a model for understanding differences in identical twins, measuring heritability of traits and the combined effects of genes and environment, with better statistics (because they are four) than with human multiple births (which are generally two).

In some tiny polyembryonic wasps, parasites of caterpillars, the embryo splits at a quite advanced stage of development, and this form of asexual reproduction eventually takes on extraordinary proportions. From a single egg deposited into a living caterpillar, very many genetically identical wasps can derive. This was first noticed by Paul Marchal (1862–1942), who in 1898 described the dissociation of the primary embryo into multiple secondary embryos in the tiny hymenopteran known today as *Ageniaspis fuscicollis*. However, the first accurate study was undertaken by the Italian entomologist Filippo Silvestri (1873–1949), who in 1906 illustrated polyembryony in a different species, known today as *Copidosoma truncatellum*, where from a single egg almost 2,000 embryos may originate. Not all of these embryos will develop into adults. The best-studied polyembryonic wasp, *Copidosoma floridanum*, can produce about 3,000 clonal descendants from a single fertilized egg that develop into two castes, reproductive and soldier larvae. Only reproductive larvae develop into adult wasps, whereas their sisters do not develop further than the larval stage and defend the reproductive members of the clone by attacking other parasites present in the same caterpillar.

Although occurring in animals that produce embryos by sexual reproduction, polyembryony is actually a form of asexual reproduction, practised by an individual still in a very early (embryonic) phase of development. The result is a set of embryos, all genetically identical to each other, but different from both parents. To be precise, what we call parents are actually the most recent sexually reproducing ancestors, since when the off-spring are more than two, more rounds of embryo splitting, and thus more asexual generations, can separate the two 'parents' from their 'offspring'. As noted in Chapter 2, when polyembryony is obligate, this reproductive mode results in a regular alternation of sexual and asexual generations, known as a metagenetic cycle.

In plants, the production of monozygotic twins by division of the original embryo within the seed is called *cleavage polyembryony*, and this is common in conifers (e.g. *Pinus*, *Tsuga*, *Cedrus*) but less frequent in flowering plants. In plants, polyembryony leads to competition for developmental resources among the embryos produced from the same ovule; the outcome, usually, is the elimination of all but one of the competitors.

Similar to polyembryony is *larval amplification*. In this case, the division into two or more genetically identical individuals occurs at the stage of larva, rather than embryo. This is limited, however, to species where a larval phase occurs, that is, when there is a juvenile phase of develop-ment that is clearly different from the adult in its morphology and/or in its ecological niche, and which generally transforms into an adult through metamorphosis. For instance, starfishes have larvae that are not star-shaped and do not live on the sea floor, but have bilateral symmetry and passively float in the water. Some species can reproduce asexually at the larval stage by paratomy at the level of one of the arms, or by self-amputation of a portion of a lobe in front of the mouth, or by budding. These diverse methods of asexual propagation provide a common mech-anism to extend the length of larval life and rapidly increase the number of individuals. Larval amplification is also found in other animal groups, including cnidarians and flatworms.

Summing up, asexual reproduction, the reproduction that leaves out sexual processes, genetically simple as it might be, can actually occur through a variety of modes (Table 3.1). In the next chapter we will move on to consider reproduction which takes sex on board.

Asexual reproduction
- in unicellular organisms
  - binary fission (either symmetric or asymmetric)
  - multiple fission
- in multicellular organisms
  - with unicellular propagules (spores)
  - with multicellular propagules
    - by accidental fragmentation followed by regeneration
    - by rhizomes, bulbs, tubers, etc. (in plants)
    - by buds, stolons, etc. (in animals)
    - cleavage polyembryony (in plants)
    - polyembryony and larval amplification (in animals)

Table 3.1. Summary of the main modes of asexual reproduction.

4 Reproduction with Sex

Gametes

In Chapter 1 we defined sexual reproduction as a form of reproduction that generates new individuals carrying a genome obtained by the association and/or the reassortment of genetic material from more than one source. In the most familiar form of sexual reproduction, the new genome is formed by the union of (partial) copies of the genomes of two parents through the fusion of two special cells produced for that purpose, the *gametes*, into a single cell, the *zygote*. This is the way most multicellular eukaryotes, ourselves included, reproduce sexually.

However, there are other ways in which sex and reproduction may be linked, as we have seen in some of the life cycles described in Chapter 2. For instance, in the unicellular green alga *Chlamydomonas*, there are cells functioning as gametes, but these are not produced by a parent organism, as for example in animals: in *Chlamydomonas*, an apparently ordinary cell (in this case, the whole organism) at a certain point of its life cycle assumes the function of a gamete.

Another variation on the theme is offered by multicellular organisms like flowering plants, in whose life cycle haploid and diploid generations alternate. Let's remind ourselves (see Chapter 1) that sexual reproduction involves two distinct sexual processes: recombination of the parent's genomes at meiosis (the products of which are gametes or spores), and the combination of the genomes of two gametes into the zygote's genome (syngamy). The combination in sequence of these two processes is what normally happens in the sexual reproduction of most animals. However, in most plants, where there is alternation of haploid and diploid generations, these two sexual processes, recombination and syngamy, belong to the reproduction of the sporophyte and the gametophyte, respectively. There is recombination in the sexual reproduction of the (diploid) sporophyte: by meiosis it produces haploid reproductive cells that are not gametes, but spores, which will develop into gametophytes. The latter, in turn, also reproduce sexually, although without passing through meiosis, and produce gametes that will fuse to form diploid zygotes, the founding cells of the next generation of sporophytes.

Another option is that the merging of two genomes into the genome of the zygote does not occur by fusion of two gametes, but rather by the fusion of two gametic nuclei that may have been present in the same cell for a long time. This happens in the self-fertilization of many protists, but also, more conspicuously, in the sexual reproduction of most fungi. As we saw in Chapter 2, the hyphae of two major fungal groups, the ascomycetes and the basidiomycetes, grow through the proliferation of a long-lasting dikaryotic mycelium (i.e. with two nuclei in each cell compartment), from which the fruiting bodies (the mushrooms) are eventually produced. In specialized structures of the mushroom, the long-delayed completion of syngamy occurs through the fusion of the two nuclei of the same cell compartment, with the production of very transient diploid zygotic nuclei that before long will produce spores by meiosis.

Sexual reproduction is found in all multicellular eukaryotes and in most protists, but not all of them. In any case its absence does not necessarily rule out other forms of sexual exchange that are not linked to reproduction, as in the peculiar dissociation of reproduction and sexuality in ciliates (Chapter 1).

When two gametes are involved in sexual reproduction, these may show different degrees of morphological and functional differentiation (Figure 4.1). In many algae, the two gametes, generally equipped with one or more flagella, are morphologically indistinguishable. That is, there are no distinct male and female gametes, and reproduction is said to be by *isogamy*. However, in many organisms with identical gametes, although these are not attributable to a specific sex, nevertheless they show a molecular marking that makes them belonging to a particular *mating type*, which allows a gamete to fuse only with gametes of a mating type other than its own.

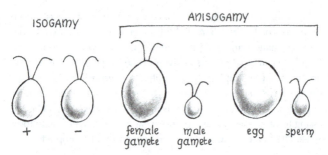

Figure 4.1. Degrees of differentiation between the gametes for a given species. Isogametes have no sex but may differ in mating type (here, +/−). The two pairs of anisogametes are examples of the two extremes in the degree of differentiation. Gametes other than the egg are depicted as bearing two flagella, as is typical of many green algae.

In very many organisms, however, two morphologically different types of gametes are found, and they are said to reproduce by *anisogamy*. Gamete dimorphism may come in different degrees. At one extreme, the two types of gametes are similar in shape and function (e.g. being both flagellated and motile), but one of the two, indicated in Figure 4.1 as the female gamete, is considerably larger than the other, indicated as the male gamete. This condition is widespread among the green algae, for example in a number of Volvocales. At the other extreme, gamete dimorphism involves both size and shape. The larger (female) gamete is immobile and generally full of nutritional reserves, while the male gamete is much smaller and usually mobile. A human egg has a volume about 100,000 times larger than a human sperm cell. The female gamete is called, according to the different taxonomic traditions, egg, ovum, ovocell, oosphere or ovule (though the last should not be confused with the ovule of seed plants). We will call it simply egg. The male gamete is called, also according to the different taxonomic traditions, sperm, spermatozoon, antherozoid, spermatozoid or spermatium. We will call it simply sperm, or sperm cell.

This apparently marginal detail concerning the exterior appearance of gametes in eukaryotes – the fact that there are distinct and male and female gametes – has enormous, far-reaching consequences for reproduction processes and their evolution. On this fundamental asymmetry is based a large part of the disparity in reproduction modes, sex roles and sexual behaviour in the tree of life.

Diversity of Gametes

Animal eggs are generally large or very large cells, owing to the accumulation of resources (yolk) that sustain the development of the embryo at a time when the animal has no organs for the intake of nourishment from outside. However, despite the almost universal presence of yolk, egg size is usually less than 1 millimetre. For example, most species of sea urchins and sea stars produce eggs with a diameter of 0.1 millimetres, while those of mammals (except the egg-laying platypus and echidna) vary from 0.07 to 0.25 millimetres. Larger eggs, however, are not uncommon in bony fishes (1–6 millimetres) and especially in rays and sharks (15–100 millimetres), and even more so in oviparous land vertebrates. Among the birds, the smallest eggs are those of the hummingbirds, the tiniest of which are just 10 millimetres in length, weighing 0.2 grams; the largest are those of the ostrich, about 16 centimetres long, with a diameter of 14 centimetres and a weight in the order of 1.6 kilograms, the biggest cell in the whole tree of life. The ostrich egg is big in absolute terms, but its weight represents only 1.7% of the animal that produces it; by contrast, there are birds where the egg has an exceptional relative weight: 13% in the Eurasian wren, 20% in kiwis and 22% in the European storm petrel.

The typical animal sperm is a small cell with an ovoid head that contains the nucleus, a midpiece containing energy-producing organelles (the mitochondria) and a tail (actually, a flagellum), often of a length in the order of 50 micrometres (μm). However, deviations from this model are frequent, especially among animals that practise internal fertilization. There are animal sperm cells with two flagella, a condition very widespread among the flatworms, while most crustaceans have tailless sperm. The sperm cells of crayfish, lobsters and crabs are tailless and non-motile 'explosion sperm': they achieve sudden short-term motility by a chemical reaction in the anterior of the sperm head that produces an abrupt eversion of the cell membrane, which causes a leap-like forward movement. Among the insects, there are sperm cells of exceptional size, up to 58 millimetres in the case of *Drosophila bifurca*, 20 times the whole animal's length. These long sperm are thought to have evolved under a form of sexual selection, where sperm from different males compete to fertilize an egg (sperm competition; see Box 4.1). Sperm of the common wood mouse (*Apodemus sylvaticus*) display a unique cooperative behaviour. The sperm head has an apical hook through which they form long 'trains' of hundreds or thousands of sperm cells, which significantly increases swimming velocity.

In several animal groups – including leeches, arthropods, cephalopods, gastropods, bivalves, and even amphibians and cyprinodontid fishes among the vertebrates – sperm are packed in *spermatophores*, which are deposited by the male on the ground or, in aquatic environments, on the seabed or on a suitable submerged support, or delivered directly to the female. The sperm are enclosed in a capsule, or agglutinated around an axile support, similar to a plume; fertilization is almost always internal. Spontaneously, or induced by the male, the female collects the spermatophore and brings it or at least its contents into contact with her genital opening.

In other animals, especially among those with internal fertilization, such as the guppy (the teleost fish *Poecilia reticulata*), sperm bundles are formed, where each bundle (or spermatozeugma) includes about 30,000 sperm cells.

Some of the sperm produced by an animal may have a function other than fertilization, such as providing nourishment to the female, or limiting the risk of competition from another male's sperm (Box 4.1).

As for the land plants, the oldest and most conservative lineages have flagellated male gametes, which reach the immobile female gametes by moving through the film of water in which a plant can be covered, at least for a while. The sperm cells of mosses and their relatives have two flagella; those of horsetails and ferns have many flagella, as have the sperm of cycads and ginkgo. In the other groups, including all flowering plants, there are no flagellated male gametes, but only cells with the value of gametes which are part of the tiny male gametophyte, the pollen grain. The female gamete of the land plants is an immobile cell, sometimes easily recognizable inside flask-shaped reproductive organs in the gametophytes of mosses and ferns. In the flowering plants, the egg cell is located at one end of the female gametophyte, and an opening through the surrounding tissues allows the passage of the two sperm cells carried by the pollen tube.

Sexes, Sex Conditions and Breeding Systems

To establish the number of sexes in a species we must first agree on what is to be counted. The simplest principle only takes into account the number of morphological types of gametes that are produced. Sex is singular (or indeterminate) in the species with gametes of one type only (reproducing by isogamy), whereas there are two sexes in species with distinct male and female gametes (reproducing by anisogamy), as in most species that reproduce sexually. End of the story? Not quite. Using the same criteria, but assimilating the notion of mating type into the concept of sex, the number of sexes in one species can reach up to several thousands, as in certain fungi.

However, simply counting the number of gamete types or mating types overlooks an essential aspect of sexual reproduction, that is, that generally a maximum of two individuals are engaged in a sexual exchange by mixing their genomes to form a new individual. From this perspective, the number of sexes would be the number of types of gametes that must be joined to form a new fertile individual in the population. This number seems never to exceed two, so that sexual reproduction basically appears to be a binary system.

Nevertheless, if we broaden our perspective, attributing a value of individual, or 'superorganism', to the level of organization of complex societies, such as those of ants, bees and other insects, the number two does not seem to be an absolute limit. In an American ant species of the genus *Pogonomyrmex*, two distinct genetic variants coexist, which we can designate as 'blue' and 'yellow', each with fertile individuals (queens and males) that produce gametes (eggs and sperm) of the respective type (blue and yellow). As in all ants, wasps and bees, females (both queen and workers) are diploid, whereas males are haploid. In a colony with a blue queen, she can generate blue males through unfertilized eggs (by parthenogenesis, Chapter 6) and new blue queens by mating with blue males. However, to generate workers for her colony the blue queen must be fertilized by male gametes of the yellow type. A reverse situation is found in the colonies with a yellow queen, where, to produce a sterile caste of workers, the contribution of blue male gametes is required. From the point of view of the colony as superorganism, the queen (blue or yellow) must mate with two types of male (blue and yellow), and must receive both types of male gamete, to generate a self-sustaining colony. A colony with a fertile queen therefore has at least three parents, descending from three types of gametes. Furthermore, four types of gametes (male and female, blue and yellow) are necessary for the perpetuation of the hybrid system.

Even with a maximum of two sexes in a species, the number of kinds of individuals with respect to their sexual function is not limited to two. This is the individual's *sex condition*, which can be either male or female, but also both male and female (hermaphrodite) or neither male nor female (sexually indeterminate). Thus, with two sexes there are four sex conditions. With anisogamy, an individual's sex condition coincides with the type of gametes it produces: male if it produces male gametes exclusively, female if it only produces female gametes, and hermaphrodite if, simultaneously or at different times, it is able to produce both types of gametes. On the contrary, with isogamy, where gametes are not distinguishable as male and female, the ascription of an individual to a specific sex condition does not apply and we could designate it as sexually indeterminate. Once more, this qualification disregards the fact that sexually indeterminate individuals may belong to a particular mating type, which allows them to have sex only with individuals with a different mating type.

In haplodiplontic organisms, with alternating haploid and diploid generations, for example land plants (Chapter 2), only the gametophyte produces gametes, and thus, strictly speaking, only gametophytes can have a sex condition. However, the notion of sex condition is normally applied to the sporophyte, despite the fact that the reproductive cells it produces are spores, rather than gametes. Thus, the sex condition of the gametophyte (male, female or hermaphrodite) is defined by the type of gametes it produces (sperm, eggs or both, respectively), while the sex condition of the sporophyte (male, female or hermaphrodite) is defined on the basis of the type of spores it produces (male, female or both, respectively). The sex of the sporophyte can also be indeterminate, if it produces a single type of spore. Gametophyte and sporophyte of the same plant do not need to have the same sex condition. Most flowering plants have hermaphrodite sporophytes (with bisexual flowers, possessing both stamens and carpels, on the same plant) and distinct male and female gametophytes (pollen grain and embryo sac, respectively; Chapter 5), but other combinations are also observed.

To these four sex conditions, at least two more should be added, those of individuals that do not produce gametes (i.e. sterile male and sterile female) belonging to a species where other individuals are fertile. In many species, this sex condition is not an anomalous condition but part of the social organization of the species. This is the case in ant, bee and wasp colonies, where the workers are all sterile females. The situation is different in termites, where, depending on the species, the sterile castes can be composed of female individuals only, male individuals only, or both. In all these cases, an individual's condition cannot be determined as male or female based on the gametes it produces; it will correspond instead to the chromosome complement (haploid or diploid, in the case of ant, bee and wasp workers; with XY or XX sex chromosomes in most termites), or to secondary sexual characters or the rudiments of the male or female gonads (termites).

While in species with isogamy there is only one sex and one sex condition (sexually indeterminate), and thus only one type of individual with respect to sex, in species with anisogamy the combination of different sex conditions that can be observed in the same species, or population, determines the so-called *breeding system* of the species. In a number of species producing two types of gametes, a single sex condition exists nonetheless, when all individuals are hermaphrodite, as in many land snails and earthworms, and the species is said to be hermaphrodite. In other species there are two sex conditions, with male and female individuals (*separate-sex* species), but also some with male and hermaphrodite individuals (the nematode *Caenorhabditis elegans*, several barnacles), or with female and hermaphrodite individuals (some sea anemones and sponges and more than 250 genera of flowering plants). Much more restricted is the number of species where three sex conditions coexist, with males, females and hermaphrodites. Examples are the common ash (*Fraxinus excelsior*), two cacti (*Opuntia robusta* and *Pachycereus pringlei*) and some crustaceans.

Summing up, with at most two kinds of gametes, we have four main types of sex conditions for an individual (male, female, hermaphrodite, sexually indeterminate) and six types of breeding systems for a species or population (separate-sex; all hermaphrodite; males and hermaphrodites; females and hermaphrodites; males, females and hermaphrodites; all sexually indeterminate). The two most common breeding systems, separate sexes and hermaphroditism, are the subjects of the next two sections.

Separate Sexes and Sexual Dimorphism

A species (or population) is said to be *separate-sex* (or *gonochoric*, in zoological terminology; *dioecious*, in botanical terminology) when it includes distinct male and female individuals exclusively. In animals, separate sexes is the norm, while in only 6% of the species of flowering plants the sporophyte (what is generally meant by 'a plant,' the generation with roots, stem(s) and leaves) is separate-sex, although this percentage is significantly higher in tropical forests and on remote oceanic islands.

In separate-sex species, the fundamental asymmetry between male and female gametes extends to the male and female individuals that produce them, who have distinct roles in sexual reproduction and, in certain species, in social interactions. This can result in the evolution of sexual dimorphism, where the two sexes (or one of the two) exhibit secondary sexual characters. These traits do not directly concern the reproductive system, but nonetheless present distinct forms in the two sexes.

Compared to animals, most separate-sex plants exhibit only modest differentiation of secondary sexual characters. Dwarf males are common in mosses, where tiny males grow upon much larger females, but the only flowering plant in which it is possible to safely identify the sex of an individual without checking the reproductive organs seems to be the hemp (*Cannabis sativa*): at least in some cultivars of this species, female plants have a larger stem, a more developed root system and larger leaves. As an aside, only the female flowers contain sufficient quantities of cannabinoids (the psychoactive components of marijuana) to be conveniently extracted. In several other separate-sex plant species, less pronounced differences between males and females can be observed in overall habitus, leaf morphology and other traits, but these are not diagnostic. For instance, a statistically significant difference in plant size is observed in yew, ginkgo and poplars, where male plants are usually taller than females.

Secondary sexual characters are much more common in animals, where their main roles are in the interactions between the two sexes (recognition of the partner, courtship), in within-sex competition for a mating partner, and in the different parental investment in the production of gametes or in the fulfilment of parental care.

Males and females of the same species can be very different, up to the point where they can easily be assumed to be different species. A sensational case of extreme sexual dimorphism was revealed in 2009 by a re-examination based on molecular characters of the relationships between representatives of two putative fish families (Megalomycteridae, Cetomimidae), which in fact turned out to be the male and the female of the same species – and larval specimens of the same species had long been placed in a third family (Mirapinnidae).

Striking characters of males can evolve as a consequence of the female's preference for such characters. For example, many male birds have a much more conspicuous plumage than the conspecific females – think of peacocks, pheasants and birds of paradise. Other conspicuous sexual characters derive from the competition between males through direct confrontation. This form of sexual selection (Box 4.1) is present for example in deer, whose males possess weapons (antlers) used in ritualized confrontations and actual fights between rivals.

Another form of sexual dimorphism re-enacts at a macroscopic level the division that exists at a microscopic level between the male gamete, small and mobile, and the female gamete, large and immobile. Females much more corpulent and less mobile than males of the same species have evolved many times independently. For instance, in many species of fireflies, the male is winged, the female is wingless. The metabolic resources spared from building and using wings and flight muscles can be used instead to produce extra eggs. In this case, the production of a large egg mass is favoured in the female by natural selection even at the cost of a progressive reduction in the animal's mobility. By contrast, in the male it seems to be more advantageous to develop the agility accompanying smaller size and those physiological and behavioural adaptations that allow males to readily find females at the time they begin to be available for mating.

In other cases, the male is larger than the female. In species where a female can be inseminated by more than one male during the same (or only) breeding season, from the point of view of a male, once insemination has occurred it is important that the female does not mate again with other males – this could frustrate the first male's reproductive effort. Multiple insemination, in fact, would trigger a sperm competition in which the sperm of the last male to have mated will most often prevail. This is as good as certain in the case of many insects, millipedes and other arthropods in which the male has copulatory organs with which he can remove the sperm deposited during previous inseminations. One of the possible strategies that allow the male to avoid this risk consists in remaining attached to the female for a sufficient time after insemination, in a position that hinders insemination attempts by other males. This behaviour is observed in several crustaceans (isopods and amphipods): here, males are considerably larger than females, contrary to the prevailing trend in the animal kingdom, and sometimes have special organs of attachment to the female's body. In most mammals, too, males are larger than females, possibly in relation to the widespread male–male competition for access to females.

Finally, sexual dimorphism can be so pronounced, that the male becomes a functionally integrated part of the female's body, opposite to the general trend that males are more active than the females of the same species. The most striking among these cases is offered by the Ceratioidei, a group of abyssal fishes related to the monkfish known as sea-devils or anglerfish. In these fishes, the male is really a dwarf in comparison to the female: in *Ceratias holboelli* the female is 60 times longer and half a million times heavier than the male! After a short independent life, the male attaches to a female, enters into intimate contact with her tissues and becomes an appendage of her body, fed by her through their conjoined circulatory systems, but still able to produce sperm. The male loses his autonomy, and from this union a chimeric hermaphrodite emerges.

In some animals, *sex-role reversal* with respect to the usual conditions is observed – for instance, in courtship (Box 4.1) or in egg incubation and parental care, as in seahorses or in Darwin's frog (Chapter 8). An extraordinary inversion of mating roles is observed in the cave insect genus *Neotrogla*, where females possess an intromittent organ (a pseudopenis), while males have no external copulatory organs but only a genital chamber (pseudovagina). The female penetrates the genital tract of the male during a long copulation, extracting sperm and the nutrient-rich seminal fluid. Females compete for access to males, so roles in sexual selection are also reversed.

Box 4.1 Sexual Selection

In evolution by *natural selection*, individuals with characteristics that raise their chances to survive and reproduce in a given environment tend to have more descendants; if these characteristics are heritable, they spread into subsequent generations, producing an adaptation of the species to its environment. But how to explain the fact that in many animals, males and females look very different, despite living in the same environment? For example, the males of many species of birds are more colourful, or have long tails, or special ornamental feathers, whereas females tend to have more cryptic plumage, matching the colours of the environment. And how to explain the evolution of characters as bizarre as the ornaments of certain males, like the peacock tail, which are of no use for survival or are even detrimental in avoiding predation? The answer is that these traits are the result of *sexual selection*. This form of selection favours those individuals that possess characters that elicit a larger reproductive success with respect to other individuals of the same sex. Sexual selection tends to favour distinct characters in the two sexes, fostering the so-called *secondary sexual characters*, phenotypic characteristics that do not directly concern the reproductive system, but nonetheless appear in distinct forms in the two sexes.

There are four main modes of sexual selection. Two of them, the most visible ones, were described by Darwin in his book *The Descent of Man, and Selection in Relation to Sex*, published in 1871. These are *female choice* and *male–male competition*. Both of them operate in advance of mating. In the former, the reproductive success, or fitness, of the male depends on the female's preference for some specific male features. Females mate with what they evaluate to be the more attractive males, based on morphological, chemical or behavioural characteristics. Common examples of preferred traits are 'ornaments' such as the long tails of the widow bird, the peacock, the swallow, but also that of many fish, like the green swordtail. In birds, amphibians and crickets, males are frequently chosen for their ability to sing. In many birds again, but also in many fishes, reptiles and butterflies, females' preferences are based on male colour patterns. The other pre-mating mode of sexual selection, male–male competition, generally takes the form of real fights or ritualized confrontations for the control of females, or for the resources needed to attract them, such as a territory, food or a nesting site. Examples of 'weapons' used in direct contests are deer antlers, or the big claw of the fiddler crabs (*Uca* spp.). The same male trait can have a double function, and be used both in male contests and to attract females, as is the case for territorial bird songs. It is generally the females who choose and the males who compete, but not always. An example of male choice is seen in the broad-nosed pipefish (*Syngnathus typhle*), where females display a temporary ornament, a striped pattern that attracts the males, who prefer ornamented females over non ornamented ones. As in seahorses, in pipefish it is the male who broods the eggs, in a ventral pouch where the female has deposited them. An example of female–female competition is provided by the jacanas. In these tropical wading birds, females are bigger and more aggressive than males, with spurs on the wings that they use in real fights to mate generally with more than one male. In this species it is the male who incubates the eggs and cares for the young. These are cases of *sex-role reversal*.

The other two main modes of sexual selection have been recognized more recently. In polyandric mating systems, where a female usually mates with more than one male (Chapter 5), a behaviour common among the insects and other arthropods and all the main groups of vertebrates, sexual selection can occur after insemination, 'hidden' in the female genital tract, or in the vicinity of the eggs, when insemination is external. This post-mating sexual selection has two forms, which parallel the two forms of pre-mating sexual selection: *cryptic female choice* and *sperm competition*. In the former, a female favours the sperm of one male among those by which she has been inseminated. This can be achieved through different mechanisms, for example by regulating the duration of the copula, by expelling the sperm of the last mate or those of the previous mates, by hindering sperm travelling through the genital tract, or by sperm storage. In species with external fertilization, fluids secreted with the eggs can help or hinder the sperm of a given male, or the choice can be exercised directly at the level of sperm–egg interactions. Cryptic female choice has been demonstrated in many species, including the guppy, a freshwater fish that has been promoted to a model organism for the study of sexual selection. Sperm competition occurs when the sperm of more than one male compete to fertilize the same eggs. Key features for the success in this competition are the number of sperm produced for each ejaculate, but also sperm velocity, longevity and vitality. Sperm competition can also include strategies to remove the sperm of other males from the female's genital tract (e.g. with the help of specialized appendages), or to create a barrier to the sperm of other males (e.g. by means of a sperm plug; Chapter 8).

A possible evolutionary upshot of sexual dimorphism and sexual selection is *sexual conflict*. This occurs when the two sexes have incompatible reproductive strategies to increase fitness, which can lead to an 'evolutionary arms race' of adaptations and counteradaptations between males and females. For instance, in many animals, males would maximize their fitness by increasing the number of matings, while multiple matings may harm or endanger females, whose reproductive success depends on the number of fertilized eggs they can produce, rather than on the number of matings. This can lead to behaviours of sexual coercion, through harassment or other strategies to force the female to copulate, and the corresponding female counteradaptations. For instance, in the water strider *Gerris gracilicornis*, females have evolved a morphology to hide their genitalia from direct, forceful access by males.

Hermaphroditism

The term 'hermaphrodite' is derived from the Greek god Hermaphroditos, son of Hermes and Aphrodite, whose body, after being merged with that of a nymph, assumed a form with both male and female features. In biology, an individual is said to be *hermaphrodite* (or *monoecious*, in botanical terminology), when at the same time or at different stages of its life it can produce both male and female gametes. Correspondingly, a species (or a population) exclusively composed of hermaphrodite individuals is said to be hermaphrodite. Hermaphroditism is the sex condition typical of some groups of invertebrates and of the sporophyte of many seed plants. Hermaphroditism is not the same as intersexuality – the latter term is used to indicate an abnormal individual of a separate-sex species with sex-specific features in an intermediate form (Chapter 7). The use of the term hermaphrodite to indicate an unusual combination of male and female secondary sexual characters in an individual belonging to a separate-sex species, including humans, is discouraged in modern biological and medical literature.

We mentioned above that in the life cycle of land plants, which involves alternation between diploid (sporophyte) and haploid (gametophyte) generations, the sex conditions in the two generations are generally different. Hermaphrodite gametophytes are found in most mosses and almost all fern species, generally associated with a sexually undetermined sporophyte (which means that it produces a single type of spore), while a hermaphrodite sporophyte is the rule in conifers (99% of species) and flowering plants (94%), always associated with a separate-sex gametophyte. The sporophyte of the angiosperms may be hermaphrodite because of the presence of bisexual flowers (with both stamens and carpels in the same flower), or because of the presence of both kinds of unisexual flowers (with either stamens or carpels) on the same plant.

Hermaphroditism is estimated to occur in 5–6% of animal species, with over 70% of animal phyla containing at least one hermaphrodite species. Leaving aside the insects, a huge group within which hermaphrodites (coexisting with true males) are essentially limited to a very few species of scale insect, almost one-third of animal species are hermaphrodite. Among the most familiar hermaphrodites are many sponges, flatworms, earthworms, leeches, terrestrial slugs and snails.

In animals, there are many forms of hermaphroditism (Table 4.1), which can be classified in two main categories: *simultaneous hermaphroditism* is when an individual matures as male and female at the same time, whereas *sequential hermaphroditism* is when male and female functions are developed at different times.

Hermaphroditism
- simultaneous (simultaneously male and female)
  - sufficient (self-fertilization possible)
  - insufficient (cross-fertilization obligate)
    - with reciprocal insemination (mutual insemination of partners during a mating)
    - with non-reciprocal insemination (each partner takes only one sexual role during a mating)
- sequential (male and female at different stages of life)
  - male-first (male first, then female)
  - female-first (female first, then male)
  - alternating (alternation between the two sexes in the course of life)

Table 4.1. Temporal and functional modes of hermaphroditism in animals.

In some simultaneous hermaphrodites, self-fertilization is possible, but not obligate. This form of hermaphroditism (said to be *sufficient*) is widespread in parasitic flatworms such as tapeworms and flukes, pulmonate gastropods and a few bony fishes – among them many groupers and the gilt-head sea bream (*Sparus aurata*). The small flatworm *Macrostomum hystrix* can resort to both cross- and self-fertilization. The latter is practised in a very unusual way: in a kind of hypodermic insemination (Chapter 5), the animal uses its very thin copulatory organ to inject sperm into the front half of its own body, from where these will swim to reach the eggs. *Kryptolebias marmoratus*, a small killifish of the mangrove environments along the Atlantic coasts from Florida to southeast Brazil, is the only vertebrate that reproduces routinely by self-fertilization. Its hermaphrodite gonad (ovotestis) produces eggs and sperm that meet in the common segment of the genital tract. Cross-fertilization occurs occasionally.

In other simultaneous hermaphrodites, cross-fertilization is obligate (their hermaphroditism is said to be *insufficient*), but the meeting of two individuals can result in different forms of mating. In earthworms, for instance, both partners act simultaneously as male and as female (reciprocal insemination). To mate, two worms pair belly to belly with the anterior part of the body, but with their heads pointing in opposite directions. Each worm ejaculates sperm that ends up in the other worm's sperm receptacle, helped by the mucous layer in which both worms are wrapped. Once sperm has been exchanged, the two partners wriggle away. Each worm produces a short slimy tube from a small set of segments called the clitellum, rich in glands, positioned about one-third of body length behind the head. This mucus tube travels forward, passes first over the sacs containing the worm's own eggs, which stick to the mucus, then over the sperm receptacle, where the other worm's sperm are kept, so that the eggs are fertilized. The mucus tube with the fertilized eggs is moved further forward, slips off the body and closes, becoming a cocoon where the worm's embryos will develop. In other insufficient hermaphrodites, leeches for example, at each exchange of gametes with a partner, an individual takes only one sexual role (non-reciprocal insemination). When two leeches mate, they line up with the anterior of the body in contact and the heads pointing in the same direction. One of the two (acting as a male) attaches a spermatophore to the other leech (acting as a female). The sperm released from the spermatophore enter the female-acting leech through its skin, eventually finding their way to the ovaries (fertilization by dermal impregnation; Chapter 5). Like the earthworm, the leech produces a tube of mucus from the clitellum that collects the fertilized eggs and protects them as a cocoon.

In sequential hermaphroditism, three modes are distinguished, depending on the sequence of sex roles during the animal's lifetime. Individuals are first male, then female in sponges, limpets and numerous shrimps. The opposite, female first, is less common and is found in some species of tapeworms and tunicates. Finally, sex can alternate in both directions, with more than one change during a lifetime. This repeatedly *alternating hermaphroditism* is the least common, and is known for some polychaetes, some bivalves (including the common oyster) and some fishes.

Sex change of the individual is often under social control. For instance, the male-first clownfishes (*Amphiprion* spp.) live in groups formed as a rule by a female, a large breeding male and several smaller males that do not reproduce as long as they remain subordinate. When the female dies or leaves the group, the breeding male becomes female and the largest among the other males takes its place as the breeding male. In some species of the genus *Lythrypnus*, tiny fishes living along the Atlantic and Pacific coasts of the American continent, bidirectional patterns of sex change have been observed. In the presence of another female, a female can be converted into male within two weeks, and the same individual can change sex many times, depending on the environmental conditions to which it is exposed, in particular the ratio of males to females in the local population.

Forms of sequential hermaphroditism are also found in plants. In *Arisaema triphyllum*, a herbaceous perennial plant native to eastern North America, known as Jack-in-the-pulpit, each year half of the individuals change sex, switching the production of unisexual flowers to the other sex, and every individual can change sex several times in the course of its life (up to 20 years or more). In *Eurya japonica*, an ornamental shrub native to eastern China, Korea and Japan, the frequency of sex change depends on the combination of physiological parameters such as rate of growth and environmental factors including the amount of light to which the plant is exposed.

Germ Cells and Reproductive Organs

In multicellular organisms, and also in unicellular organisms with advanced colonial organization such as *Volvox* and other green algae, there is a distinction between *germ cells*, which are the potential sources of the reproductive cells (gametes or spores), and thus of the next generation, and somatic cells, which will not bring any genetic contribution to the next generation. In haplontic and diplontic organisms, and in the gametophyte of haplodiplonts, the germ cells give rise to the gametes, while in the sporophyte of haplodiplontic organisms and in the fungi the germ cells produce spores.

This distinction between germ cells and somatic cells was introduced by August Weismann (1834–1914) in 1892. In his germ-plasm theory, Weismann presupposes a sort of immortality of a 'germ plasma', present only in the germ cells in the gonads (those from which egg and sperm develop), but not in any of the other (somatic) cells of the body. Heritable information cannot pass from germ cells to somatic cells, so that only the former are involved in the transmission of characters across generations.

Germ cells often occupy a distinct position within the body (or colony). At the time the individual reaches reproductive maturity, the production of gametes (*gametogenesis*) or spores (*sporogenesis*) is commonly carried out in specialized body regions or organs, different from group to group, called *reproductive organs*. However, this anatomical observation does not equate to the often taken-for-granted notion that germ and somatic cell lines separate (segregate) very early in development.

An early segregation of the germ cells in the course of development is found only in some groups of animals and, in general, not in plants. Moreover, regardless of the early or late segregation of germ cells, in some animal groups, such as the sponges, gametogenesis is diffuse across the body. A lack of true reproductive organs is often coupled with a lack of specialized pathways for the release of gametes, which occurs instead through ducts and openings with a different function (e.g. excretory, in some polychaetes), or by simply breaking through the body wall.

Most animals, however, have distinct reproductive organs, collectively part of the *reproductive system*. These are essentially the *gonad* (female gonad or *ovary* and male gonad or *testis*) where germ cells are found accompanied by specialized somatic tissues, a canal (*oviduct* and *deferent* duct, respectively) and an opening (*genital pore* or *gonopore*) through which the gametes reach the outside world. In the case of hermaphrodite animals, male and female structures are usually distinct, but in some groups (molluscs, some fishes), there is only one gonad (*ovotestis*) with both female and male functions.

Despite the simplicity of this schematic description of the core structure of the reproductive systems of animals, in practice these are almost always much more complex. In addition to producing gametes and releasing them outside, the animal can do many other things to maximise its reproductive success. Additional functions need additional organs: let's briefly review the main ones, considering separately the organs and functions of the female and those of the male (or of the female versus the male part of the reproductive system of a hermaphrodite).

To begin with, the egg is a special cell not only because it is a gamete (a haploid cell that unites with another haploid cell, a sperm, to give a zygote), but also because in its cytoplasm are stored nutritional resources (yolk) that will sustain the embryo for a time. At least, this is the most frequent condition, which in any case requires an ovary structure capable of supplying yolk to the cell that is maturing as an egg, a supply that very often involves the complete reabsorption of the vitelline cells in which the yolk molecules are synthesized. But in tapeworms and many other flatworms, both parasitic and free-living ones (e.g. planarians), the vitelline cells are not in the ovary, but are produced in an independent organ (vitellarium); the animal eventually releases compound eggs, in each of which the female gamete, produced as usual in the ovary, is covered by a layer of vitelline cells filled with yolk. The content of these cells is not transferred to the egg, but will be used later by the embryo.

Other structures of the female reproductive system are typical of animals that practise internal fertilization. These may include a vagina designed to accommodate the partner's intromitting organ and a spermatheca in which the sperm are stored until fertilization, which does not necessarily occur immediately after insemination, for a number of possible reasons, including the fact that the inseminated female may not have, at the time, mature eggs ready for fertilization.

In some animals, including almost all mammals, the eggs fertilized inside the female's body are not laid soon after fertilization, but are hosted by her for a while. A further level of complexity emerges when close interactions develop between mother and offspring, with the appearance of structures suitable for the nutrition of embryos, as in the case of the placenta of mammals. More on this in Chapter 8.

In the male, the structures that complete the reproductive system are mainly represented by the copulatory organ, which is most often single, but is sometimes paired (e.g. in mayflies, crabs and sharks). However, the condition of the polychaete worm *Pisione remota*, which has a few dozen of them, is definitely exceptional. Beyond the obvious function of sperm delivery, a copulatory organ can also be loaded with other functions, for example to remove the sperm of competing males from the female genital tract (e.g. the sawfly *Athalia rosae*) or even to damage the female genital tract in order to prevent additional matings with other males (e.g. the beetle *Callosobruchus maculatus*).

In embryophytes, and plants in general, the gametes are produced by the gametophyte in structures that have the value of reproductive organs, called *gametangia*. The gametophyte can be hermaphrodite, that is, carry both male and female gametangia, or separate-sex, thus producing a single type of gamete.

In land plants gametangia are always multicellular and the germ cells (which will give rise to the gametes) are always surrounded by a coating of sterile cells (analogous to the somatic cells of animal gonads) with a protective function. In land plants with more conspicuous gametophytes (mosses, ferns and relatives), these represent male and/or female gametangia (*antheridia* and *archegonia*, respectively). By contrast, in seed plants, where the gametophytes are very small, it is the whole gametophyte that takes on the role of a reproductive organ.

The male gametophyte includes only the few cells of a pollen grain. This has a different structure in the different groups. In flowering plants, within a pollen grain we generally recognize a vegetative cell (also called the tube cell, because it will produce the pollen tube) and a generative cell which by mitosis forms two sperm cells, one destined to fertilize the egg, forming a zygote, while the other fertilizes another cell of the same gametophyte (the central cell), from which the nutritive tissue of the seed (the secondary endosperm) will develop (double fertilization; Chapter 5).

The structure of the female gametophyte of the seed plants differs even more extensively from group to group than does the male gametophyte. In gymnosperms, it retains for a long time a coenocytic organization; in other words, it contains many nuclei (up to several thousand) within a common cytoplasmic mass, to subsequently develop several egg cells that can be fertilized (usually 2–4, although only one zygote will develop into an embryo). In the flowering plants the female gametophyte is called the embryo sac, and is formed by seven cells only, one of which is the egg cell.

Spores are produced by the sporophyte in structures called *sporangia*. In species with separate male and female spores, in the female sporangia, germ cells undergo meiosis, producing four haploid spores, of which only one will become a female spore, while the others degenerate. In the male sporangia, each germ cell produces, by meiosis, four haploid male spores. Sporangia can in turn be borne on specialized structures, which are analogous to animal reproductive systems. These are called sporangiophores (e.g. in horsetails) or sporophylls (e.g. in ferns and seed plants). Still more complex structures are the strobili of gymnosperms (e.g. pine cones) and the flowers of angiosperms.

In seed plants, the complex of a female sporangium plus the nearest tissues of the sporophyte that surrounds it is called an *ovule*. Following the development of a spore into a female gametophyte and the subsequent fertilization of the egg cell contained in the latter, the ovule develops into a seed. In gymnosperms, ovules are borne on the surface of the sporophylls, while in angiosperms they are instead enclosed in an ovary formed by the female sporophylls (carpels).

In angiosperms, the reproductive structures, grouped in the flower, are male and female sporophylls (stamens and carpels, respectively). In a typical stamen, a proximal sterile portion (the filament) is distinguished from a distal fertile portion (anthers) that contains the mother cells of the spores. The carpels of a flower are generally fused to form a pistil, whose dilated lower part (the ovary) houses the mother cells of the female spores. The upper part of the pistil includes the style and the terminal stigma, the structure which receives the pollen. In hermaphrodite angiosperms, the flower can be bisexual, bearing both stamens and carpels, or unisexual, carrying only male organs or only female organs. Separate-sex flowering plants clearly have only unisexual flowers.

The Evolutionary Enigma of Sex

The oldest fossil record of sexual reproduction are the gametes of the red alga *Bangiomorpha pubescens*, dated about 1.2 billion years ago. Even so long ago, sex was arguably not a novelty. Comparative studies of the genomes of a wide range of animals, fungi, plants and protists have revealed that all major eukaryote lineages share a set of near-universal genes involved in the process of meiosis. This suggests that sexual reproduction is likely to have evolved with the earliest eukaryotes, about two billion years ago. However, even if sexual reproduction is widespread in all major eukaryotic groups, it seems to present an insurmountable disadvantage compared to asexual reproduction. The problem of the origin and maintenance of sexual reproduction is considered by many as the main problem of evolutionary biology, often labelled as the 'paradox of sex'.

With the same reproductive investment (number of eggs), females that reproduce asexually can have twice as many grandchildren (second-generation descendants) as females that reproduce sexually, simply because they do not waste resources by generating males, which do not produce offspring by themselves. This is the so-called 'twofold cost of sex', but more correctly it should be called the 'cost of males', because it does not emerge in cases where there is sex but there are no males, as for example in simultaneous hermaphrodite species. Moreover, sexual reproduction seems also to be a risky strategy for maintaining favourable gene combinations that had been stabilized by selection in previous generations; with its random remixing of existing genes association in the genome, it will easily create deleterious or non-viable combinations of genes. Nevertheless, reproduction through separate sexes has evolved repeatedly, and independently, in many eukaryotic clades.

Given these heavy costs of sex, it is fair to hypothesize that sexual reproduction must provide some selective advantage, to an extent that at least compensates for the disadvantages. Many hypotheses have been formulated, generally based on the idea that in the face of an undeniable deficit in terms of number of descendants, sexual reproduction can lead to an improvement in the 'quality' of offspring in sexual populations. Most of these hypotheses consider that sex facilitates adaptation to new environments by combining favourable genetic variants from different genomes, or that sex maintains adaptation by removing deleterious mutations more effectively, or by dissociating them from beneficial mutations in the genomes where these appear. Different types of advantages could obviously operate in a synergistic way. Recent theoretical work suggests that occasional or conditional sex, involving facultative switching between sexual and asexual reproduction, is often the optimal reproductive strategy. Therefore, the 'paradox of sex' should be reformulated as the 'paradox of obligate sex'.

However, if sex must be accepted as an evolutionarily winning strategy, a new, almost opposite problem emerges: 'how do some organisms manage without sex?' The question arises from the discovery of a number of multicellular lineages, plants, fungi and animals alike, that apparently have not been reproducing sexually for millions of years. If the prevalence of sexual reproduction shows that it must necessarily have some advantages over asexual reproduction, either those thus far hypothesized or others, how is it possible that a few groups of organisms have survived millions of years (and millions of generations) by reproducing strictly asexually? Among the animal examples of 'ancient asexual scandals', all small-size invertebrates, which more precisely reproduce by ameiotic parthenogenesis (Chapter 6), there are a few families of oribatid mites, the darwinulid ostracods (a group of freshwater crustaceans) and the bdelloid rotifers, also aquatic, but able to survive desiccation in a peculiar condition of 'suspended life' known as anhydrobiosis. In the latter, horizontally acquired genes (Chapter 1) and several forms of genetic reshuffling have been suggested to provide at least some of the benefits of sex. But evidence is not conclusive, and the enigma of sex lives on.

Summing up, in this chapter we have seen that gametes determine both the sex condition of the individual that produces them and the breeding system of the population, or species (Table 4.2). Gamete dimorphism brings with it enormous consequences for the features of males and females and their roles in the pair, in the society or in the species. In the next chapter we will turn to sexual reproduction 'par excellence', in the form of reproduction with two parents.

Sex condition of an individual
- male (produces sperm)
- female (produces eggs)
- hermaphrodite (produces sperm and eggs)
- sexually undetermined (produces isogametes)

Breeding system for a population or species
- separate-sex
- hermaphrodite
- with males and hermaphrodites
- with females and hermaphrodites
- with males, females and hermaphrodites
- with isogamy

Table 4.2. Summary of the conditions with respect to sex for individuals, population or species.

5 Two-Parent Sexual Reproduction

The Meeting Place of Gametes

For most people, the most obvious thought or image that sexual reproduction brings to mind is that of sexual intercourse, a mating between two individuals of opposite sexes, which will result in the birth of their common offspring. While biparental reproduction is certainly the most common mode of sexual reproduction among all eukaryotes, it is not the only one, and the way it is carried out can depart substantially, in many different ways, from the 'canonical' description above. What is common to all these modes is that two distinct sexually compatible individuals (parents) undertake a sexual exchange that leads to the generation of new individuals with a genetic constitution obtained from the association and/or the reassortment of those parents' genomes. The key event in this mode of reproduction, technically called *amphigony*, is the fusion of two gametes or two nuclei functioning as gametes (*syngamy*), each produced by one parent, to form a zygote. While in species with anisogamy (i.e. with distinct male and female gametes; Chapter 4), only gametes of opposite sex are compatible, the two individuals that produce them are not necessarily a male and a female. Rather, other combinations are also possible: a male and a hermaphrodite providing the egg (as in the nematode *Caenorhabditis elegans*); a female and a hermaphrodite providing the sperm (as in the shrimp *Pandalus borealis*); or two hermaphrodites, one providing the egg, the other the sperm (as in the mating of two leeches). In fungi and other eukaryotes with isogametes, the two parents must belong to different and compatible mating types. However, beyond these variations in the sex condition of the parents, there is enormous diversity in the ways the two gametes meet and merge, as we will see in the following pages.

Release and/or transfer of sperm is not the same as *fertilization*, which strictly speaking coincides with syngamy, the merging of male and female gametes. The release of gametes to the external environment, either by both sexes or by the male only, is known as *spawning*, whereas, when male gametes are released in the immediate vicinity of the eggs, whether inside the body of the female or externally, the process is called *insemination*.

External fertilization means that gametes meet freely in the environment, generally in water or another fluid where, in most instances, at least the male gametes can swim. In the unicellular parasite *Plasmodium*, fertilization occurs in the intestine of the intermediate host, a mosquito, filled with blood sucked from a warm-blooded vertebrate. External fertilization can be achieved in different ways: most often, and most simply, through the spawning in the water of gametes of both sexes, as in many algae and marine invertebrates, or by *external insemination*. The latter refers to a situation where the male releases sperm directly over the eggs just laid by the female, as in most frogs and toads and a number of fish species. A special case of external insemination occurs when sperm are transferred by means of spermatophores (Chapter 4) and the female receives them not in her genital tract but in another receptacle. For example, in symphylans, a group of small, delicate soil arthropods, the male deposits stalked sperm drops on the soil, in the absence of a female. The female bites off the sperm drops and stores them in her mouth. The eggs are smeared with sperm at the time they are deposited on the ground. Thus, although insemination is typically a male job, in this case the last step in insemination is actually brought about by the female.

Internal fertilization means instead that the male gamete merges with the egg while the latter is still inside the body of the female, or protected by her in body folds or cavities. In separate-sex animals and in those with insufficient hermaphroditism, internal fertilization presupposes the transfer of sperm from the male to the female. This usually involves active *internal insemination*, or the production of spermatophores (*external insemination*). In some orthonectids (tiny animal parasites of some marine invertebrates), the whole male enters the female during fertilization. In many cases, however, internal fertilization is accomplished by the release into the external environment of male gametes that individually swim to reach the egg cells retained by the females. This form of fertilization, combining sperm-only spawning and internal fertilization, is also called *in-situ fertilization*. This is practised by many algae, mosses and ferns, and by sessile aquatic animals such as many sponges, corals, entoprocts and bryozoans.

The aforementioned categories do not apply to fungi and unicellular eukaryotes that nonetheless have amphigonic reproduction. Strictly speaking, unicellular organisms do not produce gametes, but transform into gametes, directly if the parent is already haploid (like *Chlamydomonas*), or by meiosis if the parent is diploid (as in *Amoeba* or diatoms). In multicellular fungi, finding a partner means that a haploid hypha grows onto another hypha of compatible mating type and fuses with it. What happens next depends on the kind of fungus. In many common moulds, each hypha forms a multinucleated structure (gametangium) with many haploid nuclei. The two gametangia merge, forming a novel structure (zygosporangium) that contains the haploid nuclei deriving from both parental hyphae. The zygosporangium develops a thick wall, thus becoming the resistant stage of the organism. Under favourable conditions, the nuclei deriving from the two parents fuse in pairs into diploid zygotic nuclei. These undergo meiosis, from which the spores of the next generation are produced, to be dispersed into the environment. As we saw in Chapter 4, in basidiomycetes and ascomycetes, to which most common mushrooms belong, upon the encounter and fusion of two haploid hyphae, a new hypha forms, with two haploid nuclei per cell compartment, which grow and produce the fruiting bodies of the fungus. Syngamy, with the formation of diploid zygotic nuclei, occurs in specific parts of the mushroom, the sporangia, which in the two groups take the name of basidia and asci, respectively. Very soon, the zygotic nuclei will undergo meiosis to produce spores.

In the following sections of this chapter, we will explore amphigony in plants and animals in more depth.

External Fertilization Strategies

Release of both male and female gametes into the water (*broadcast spawning*), followed by external fertilization, is a primitive strategy, simple but potentially costly. Contrivances that reduce costs by increasing the chances of meeting between gametes have independently evolved many times.

In many algae and marine animals that resort to external fertilization, the meeting of gametes is promoted by the spatial distribution of adults, which is frequently aggregated, as in the case of sponges, corals and sessile molluscs, but also by the strict synchrony with which the whole population releases gametes into the water. In the algae, release of gametes can be induced by different factors such as photoperiod or light intensity above a certain threshold. The same factors are the environmental cues used by flowering plants to synchronize flowering among the individuals of a species, which may result in more effective pollination (see below in this chapter). In the separate-sex *Monostroma angicava*, a marine green alga living on the Pacific coasts of the Japanese island of Hokkaido, gametes are released every two weeks from February to June, during the low daytime tide of the first day of new moon or full moon. In contrast, a link with the lunar or tidal cycle does not seem to exist in the case of numerous green algae of the Caribbean, in which there is a massive release of gametes before dawn, with close synchrony between the individuals of the same species (the male gametes, however, are often emitted a few minutes before the female ones) and a clear temporal separation between different but related species. Among marine animals, a well-known example is the periodic swarming of a polychaete, the palolo (*Palola viridis*), which takes place in the waters around Samoa during the nights of the second or third day after the third quarter moon of October or November. At short range, the encounter in water between gametes of opposite sexes is often facilitated by chemical communication between these cells, with the emission of attraction substances (pheromones) acting as a booster. This form of communication is widely used by animal gametes, and also by the mature individuals that produce them, as described more fully in the following section.

In some animals, the male releases sperm directly over the eggs just laid by the female while seizing her, in a behaviour called *amplexus*. This is typical of most anurans (frogs and toads) and certain crustaceans (isopods). In anurans, the male generally grasps the female from behind with his forelimbs; however, as with other reproductive traits in anurans, amplexus position is very diverse and species-specific. A male may clasp the female with his forelimbs around her waist (inguinal), under her forelimbs (axillary), around her throat (gular) or pressing under her jaw (cephalic); otherwise, the male can simply sit on the back of the female, allowing his sperm to run onto her back – and this is only a small sample from the repertoire of the chaste (there is no copulation) Kamasutra of anurans. The diversity of amplexus techniques is related to sexual size dimorphism, the type of parental care (if any) provided by the species, and a wide range of ecological factors affecting the site of oviposition.

Many fishes (but also horseshoe crabs) build or excavate nests where eggs are laid and fertilized, thus limiting gamete dispersal, in addition to offering protection to the offspring during early development. After fertilization, the brood may or may not be guarded. The male stickleback, a small freshwater fish, builds a nest using vegetal material glued by a secretion he produces. He then attracts a female to the nest, and through a complex courtship induces her to lay her eggs inside the nest, where they are inseminated. The male guards the eggs until they hatch and even beyond. As an aside, this is also an example of sex-role reversal, with the male both exerting mate choice and ensuring parental care (Chapter 4). Conversely, for instance, in salmon and their relatives, including trout, the female digs a nest in the gravel or sand of the riverbed. Then she lays the eggs, which are immediately inseminated by the male. After that, the female buries the brood, and the nest is abandoned.

Internal Fertilization: Gamete Encounters in Animals

The ways in which gametes come into contact differ very much from group to group, and in many cases even between closely related species.

In aquatic animals, in particular among those, like sponges, corals and other invertebrates, that do not move or do not move much, sperm-only spawning (*free spawning*) is often combined with internal fertilization (*in-situ fertilization*). Sperm reach the eggs by freely swimming in the water, possibly guided by chemical signals emitted by the eggs or by the female reproductive organs.

Much more common is *copulation*, where the male releases the ejaculate directly into the female structure suitable for receiving sperm. The male usually has an intromittent organ (copulatory organ), but copulation may also consist of the simple juxtaposition of the genital openings of the two partners, as happens in most birds, where males have no penis (exceptions are ducks, geese, swans and large flightless birds such as ostriches and emus). Devices to facilitate sperm expulsion (e.g. muscular pumps) are present in the penis of mammals, as well as in the copulatory organs of insects (aedeagus) and spiders (pedipalps). In comparison, spermatophores may seem to represent only a passive method of sperm delivery, but this would not be correct. Spermatophores of scorpions and some pseudoscorpions contain mechanical devices comparable to levers, and those of other pseudoscorpions, crickets and cyprinodontid fishes use osmotic systems of sperm release. Spermatophores of ticks open in a very special way: when the male inserts the neck of the spermatophore into the genital tract of the female, a chemical reaction takes place which produces CO_2, causing the explosive expulsion of the content.

In some animal groups, the copulatory organ is located at a considerable distance from the male genital opening. In these cases, before copulation it is necessary for the male to transfer the sperm from his genital opening to the copulatory organ. This is the case in dragonflies and damselflies, spiders and cephalopods. In dragonflies and damselflies, the male copulatory organ is located on the ventral side of the second abdominal segment, while the genital pore opens on the ventral face of the ninth segment (the same position at which the female genital pore is located). This peculiar anatomical arrangement explains the curious positions assumed by the two partners during mating: the male clings to the female by grasping the first segment of her thorax, using his terminal abdominal appendages. The female, meanwhile, flips her abdomen forward, eventually bringing her genital opening into contact with the copulatory organ of the male: the result is the characteristic heart-shaped posture of the two mating partners.

In spiders, the copulatory function is entrusted to the male pedipalps, the second anterior-most pair of appendages, often very profoundly modified. In two genera of Theridiidae, the subadult male amputates one of his two pedipalps, so that as an adult he possesses only one copulatory organ. During mating, the female tears the pedipalp from the body of the male, and then eats the rest of him. The torn pedipalp remains inserted for hours in the female's genital tract, continuing to release its sperm load and, most likely, preventing sperm competition from other males.

In some animals, the male copulatory organ is not introduced into the female's genital tract, but is used instead for penetrating through her body wall, frequently in a specialized area. This is insemination by *hypodermic* (or *traumatic*) *injection*. In the females of the bed bug and relatives, this special-ized area has the dual function of limiting damage from traumatic insemin-ation, being covered by cuticle with lower resistance to penetration and which heals more easily, and of intercepting the path of the sperm to the eggs to be fertilized, in the meantime converting a considerable amount of sperm into a source of nourishment.

Similar to insemination by hypodermic injection is insemination by *dermal impregnation*. Sperm deposited on the body surface of the female penetrate independently into the body of the latter. This mechanism is known, for instance, in many marine annelid worms (polychaetes), arrow worms (chaethognaths), velvet worms (onychophorans) and leeches, as mentioned in Chapter 4.

By means of *spermatophores*, internal fertilization is combined with external insemination. Among terrestrial animals, spermatophores may be released randomly, hoping for females to stumble across them by chance, or deposited close to a female, after chemical or tactile interactions with her. In scorpions, a male courts a female and directs her towards the spermatophore he has just released. Among aquatic animals, where more options for sperm movement are available, spermatophores are common among the crustaceans. In most cephalopods, one arm, sometimes greatly modified, is specialized to store and transfer spermatophores to the female. In some species it ends up detached from the male's body, remaining attached to the female. The famous French zoologist Georges Cuvier (1769–1832) misinterpreted one of these arms that he found embedded in the mantle of a female paper nautilus (*Argonauta argo*) as a parasitic worm and gave it the scientific name of *Hectocotylus* (meaning 'hundred little cups', i.e. suckers, from more or less corrupted Greek words) – hence the modern anatomical term for this specialized arm, hectocotylus.

In a number of separate-sex animals with very strong sexual dimorphism, the two partners become joined in a permanent union. In some cases, the male is larger than the female, and hosts his companion within a groove, as in the digeneans of the genus *Schistosoma*, parasitic flatworms infecting humans (interestingly, this is one of the very few non-hermaphroditic flatworm gen-era). More common, however, are dwarf males living inside the female's body, as in the echiurid annelid *Bonellia viridis*, where the female is 500 times longer than the male (Chapter 7). In the spionid polychaete *Scolelepis laonicola*, the dwarf male also lives as a parasite inside the female, but in this case the epidermis and cuticle of the two partners are continuous in the contact zone and anastomoses are formed between their blood vessels, similarly to the ceratioid fishes we met in the previous chapter.

From the moment of insemination, the life span of a sperm cell is often short, a few hours to a few days (as in most mammals), but in some animals the male gametes remain viable much longer inside the female's spermatheca, protected by the secretions of specialized glands. The life span of these sperm is in the order of a few weeks in many birds and insects, several months in many snails and in some fishes and salamanders, up to several years in a few snakes and turtles. Consequently, a single insemination allows the fertilization of eggs that will mature over a long time: in bees and ants, for example, the sperm stored in the course of the nuptial flight will remain viable for the entire life of the queen, up to five years in the honey bee.

Internal Fertilization: Gamete Encounters in Plants

Among the plants, bryophytes and pteridophytes practise a form of in-situ fertilization, with male flagellated gametes reaching the immobile female gametes by swimming in a film of water, drawn by a chemical attractant produced by the female gametophyte. Morning dew is sufficient water for the sperm to swim. Interestingly, recent experimental studies have shown that in some moss species the probability and effective distance of cross-fertilization are increased by small soil arthropods such as springtails and mites. The moss *Ceratodon purpureus* even emits volatile substances capable of attracting different species of small arthropods. These, moving through the moss carpet, act as sperm transport vectors, thus playing in cross-fertilization a role similar to the pollinators of the zoophilous seed plants (see below).

In the latter, the female gamete is an immobile cell, solidly integrated into the female gametophyte. Furthermore, in all angiosperms and in most gymnosperms the male gamete is led to fertilize the egg cell through the pollen tube developing from the male gametophyte. A free, mobile male gamete is found only in a few gymnosperm groups (cycads and ginkgo). In *Cycas*, for example, the male gametophyte (a pollen grain) is carried by the wind until it comes into contact with the ovule, which generally carries two huge egg cells. Here the

male gametophyte develops a pollen tube that contacts the female gametophyte, absorbs nourishment from it and then dissolves, releasing two flagellated sperm cells that reach the egg cell by swimming. Generally, only one of
the two embryos that may be formed from two distinct fertilization events in
the same ovule completes development into seed.

A peculiar process of *double fertilization* has evolved in flowering plants. This
consists in the fertilization, by two sperm cells carried by the same pollen tube,
of two distinct cells of the same ovule: the egg cell and the central cell. The
diploid zygote obtained from the fertilization of the egg cell, which contains one
haploid nucleus, will develop into the embryo of a new sporophyte. From the
fertilization of the central cell, which is binucleate, and from the subsequent
fusion of the three haploid nuclei, a triploid tissue will develop, called the
secondary endosperm: it will provide nourishment to the embryo (Chapter 8).

The completion of male gametophyte development is often postponed until
after its arrival on the female gametophyte or on a receptive structure such as
the stigma of a flower, from which the pollen tube can be extended towards an
ovule. The time interval between pollination and fertilization varies conspicuously among seed plants. In gymnosperms, it ranges from a few weeks (as in
most cypresses and firs) up to more than one year in *Pinus* and some araucar-
ias. In angiosperms, the interval between pollination and fertilization can be
as short as 15–45 minutes (e.g. in dandelions) or as long as 12–14 months in
certain species of oak.

The diversity of ways by which the male gametes are transferred to the female
reproductive organs corresponds to a diversity of pollination modes by which
the pollen grain is brought close to the female gametophyte. Strictly speaking,
this process amounts to an encounter between two sexual partners – the two
gametophytes of opposite sex – rather than two gametes, so we will treat it as
such in a later section of this chapter.

Partner Encounters in Animals

In animals with internal fertilization, reproduction is preceded or accompanied by various movements, or by specific forms of communication between
individuals of opposite sexes, or both. The two kinds of actions are often

associated, for instance, in the active search for a partner, where the movements of (usually) male individuals are guided by attraction signals emitted by the females. These can culminate in forms of courtship, sometimes complex and more or less ritualized. The behaviours involved are generally mediated by various types of signal, such as chemicals or sounds. In addition to serving as a call that facilitates meeting between conspecifics of opposite sexes, these forms of communication are important mechanisms of precopulatory reproductive isolation between closely related species, because of the specificity of the signals.

In order to reproduce, animals may undertake movements as adults, generally covering a short distance until a partner is encountered. For sessile species and those that move very little as adults, especially in the sea, dispersal of juvenile stages be seen as movements in preparation for adult reproduction. But this occurs also on land – for instance, in the females of scale insects, which, after a first motile post-embryonic stage, become irreversibly sessile.

However, reproduction often involves long-distance journeys. These usually have the character of seasonal migration from a region where the reproductive season is spent, to another region that offers better trophic conditions during another part of the year. In short-lived animals, these two journeys are performed by individuals of successive generations, as in the monarch butterfly. In addition to a number of sedentary or almost sedentary populations, this species includes populations that perform regular cycles involving up to four generations per year: the adults of the last generation of the year migrate south (up to 3,000 kilometres) to the wintering area; at the beginning of the new year they reproduce, starting a series of two or three generations that follow each other during the summer. These generations move stepwise as adult butterflies to the north, to finally give life to the new generation that will migrate south before the next winter.

In animals in which life expectancy is more than one year, one individual takes both trips. In the Atlantic salmon, which spends most of its life in the sea but reproduces in the fresh and oxygen-rich waters of mountain streams, a few years pass between the descent to the sea as an immature and the ascent of freshwater rivers where the adults will reproduce. Long journeys in the opposite direction characterize the migrations of eels, which are born in the sea but spend a few years in freshwater environments before returning to the sea. In these fishes, the individual performs only one migratory cycle in its life.

In other animals whose life spans several years, however, the individual travels more than once between the breeding area and what is generally defined as a wintering area. This is the case in many birds, some of which fly long distances every year. The record holder is the Arctic tern, which in the northern summer lives in the Arctic and Subarctic regions of North America and Eurasia, while it spends the austral summer on the coasts of the Antarctic continent. It has been estimated that these birds fly over 70,000 kilometres each year, with a documented case of 91,000 kilometres, the longest migration so far recorded for any animal.

Some animals that cannot be called social in the ordinary sense of the term nevertheless form temporary aggregations preparatory to mating. Examples are the swarms (males only) of mayflies and chironomid (non-biting) midges that often form over the water bodies where they lived up to metamorphosis. Less easy to observe are the aggregations of both males and females of some bark beetles, such as *Dendroctonus ponderosae*, which are attracted by a mix of the trees' resins and their own pheromones to the trunks of the conifers from which they emerged, in places that now become mating arenas.

In animals where adults are not social, or at least gregarious, the forms of communication that allow the identification of a partner of their own species and of the opposite sex are very important, and sometimes involve long-distance calls or dialogues. Communication can occur through a range of different sensory channels.

Vocal communication, widespread among terrestrial vertebrates, is also well known in some groups of insects such as crickets, grasshoppers and cicadas, but it is actually much more common than our ears can perceive.

The widespread *chemical communication* involves the release of attraction pheromones (or sex pheromones) by the female (or, in a few cases, the male) and the presence in the male of specialized receptors, often sensitive to single pheromone molecules. In the male of the silkworm moth (*Bombyx* spp.), on the antennae of which there are 17,000 chemical receptors sensitive to bombykol, to trigger the insect's search reaction, it is sufficient that one receptor in a hundred be stimulated by a single molecule of the pheromone. Sex pheromones are very diverse, but are generally characterized by a relatively low molecular weight, which represents an acceptable compromise between two opposite needs: the specificity necessary to maintain the message at the level of communication between conspecifics, while also reducing the probability that it becomes an easy attractive cue for a predator or a parasite; and the ease of dispersal through water or air, necessary to carry the message far enough to guarantee a good chance of meeting with a male, or a male gamete. Some attraction pheromones are made up of a single substance (such as bombykol, the sex pheromone of the silkworm), but very often they are mixtures of a number of different molecules.

Communication by *light flashes* is practised by fireflies. In these insects (which are beetles, not flies), the males are always winged and fly when in search of a partner. Females are winged in some genera (even if less active and mobile than males), but are often larviform and remain on the ground or, at most, on low herbs and grasses. Both sexes have an organ that allows the emission of flashes of light; dialogues between them allow both the localization of a prospective partner and the identification through a sort of Morse code of the species and sex of the other individual. Some deep-sea fishes (e.g. lanternfishes, dragonfishes) possess species-specific bioluminescent structures, sometimes sexually dimorphic, that are used in mate recognition.

Finally, *tactile communication* is an important part of the mating behaviour of many species. It can be part of species and/or sex recognition mechanisms (as in spiders), it can be used during courtship (as is common among mammals), or it may function as a way to coordinate copulatory behaviour (as in several octopus species).

Partner Encounters in Seed Plants: Pollination

In seed plants, the meeting of gametes can only be achieved if the male (pollen grain) and female (embryo sac) gametophytes are brought in close proximity, a condition that often occurs after a pollen grain's long journey. This is *pollination*, and it formally can be considered equivalent to the encounter of two sexual partners (two gametophytes of opposite sex) rather than to fertilization. These encounters come about thanks to three main carriers: wind, water and animals.

Having pollen that is carried by the wind (*anemophilous pollination*) is common among the gymnosperms (only gnetaleans and cycads are insect-pollinated, the latter typically by thrips), but also by no means rare among the angiosperms, in particular among the monocots (e.g. grasses). This is perhaps the most primitive form of pollen transport. Since wind is not a specialized vector, anemophilous plants entrust the success of pollination to abundant pollen production. Many species have evolved specific morphological adaptations that facilitate the dispersal of pollen and increase the probability that this is intercepted by the reproductive organs of conspecific plants. In the flowers of anemophilous angiosperms, a large quantity of pollen grains is produced in anthers carried by long flexible filaments, which increase pollen exposure to wind, while the stigmas are often long and feathered, a shape that allows easy pollen capture. Pollen grains of anemophilous species are light, small and sometimes equipped with devices, such as the air sacs in conifer pollen grains, which favour suspension in the air. The flowers of anemophilous plants are generally small and inconspicuous, lacking attractions for pollinators such as coloured petals or nectaries.

Pollination mediated by water (*hydrophilous pollination*) is limited to some freshwater (e.g. *Najas*) and marine plants (e.g. the seagrass *Posidonia*). These are not algae, but flowering plants with roots, stems, leaves and flowers that live under water. The pollen is carried in the body of water (i.e. under the surface) by the stream current or by sea currents, respectively. In contrast, in brackish-water species of ditchgrass (*Ruppia*), the pollen is transported on the water surface. There are also forms of 'water-assisted' pollination, different from true hydrophily. For example, in the eelgrass *Vallisneria spiralis*, a separate-sex species of slow-moving waters, male flowers develop within a leaf-like structure (spathe) that opens at maturity. The flowers detached from the spathe are carried on the water surface and, like minuscule boats, float with the anthers upwards until they come into contact with female flowers: these also open on the surface, but remain attached to the plant. The pollen grains leave the anthers when the male flower bumps into a female flower, so the pollen actually runs along the anther-to-stigma route above the water surface. Other aquatic plants with flowers floating on the water surface, such as water lilies (*Nymphaea*), rely on insect pollination.

The transfer of pollen can be mediated by the action of an animal (*zoophilous pollination*). There are a number of animal species that act as pollinators: insects, especially moths, butterflies and bees, but also true flies and beetles, birds (hummingbirds and some parrot species in the Americas, honeyeaters in Australia, sunbirds mainly in Africa but also in southern Asia and Australasia), bats and snails. Other animals can occasionally act as pollinators. Among these are some nocturnal primates and some species of opossum, which open the flowers in search of nectar, thus getting their hair dusted with pollen, and some lizards, geckos and skinks, whose muzzles become smeared with pollen when they lap the nectar.

The diversity of pollinators that ensure pollination of different plant species is generally reflected in specific flower traits evolved as adaptations that facilitate recognition by the potential pollinator, sometimes at a distance, or the transfer of pollen from the stamens of a flower to the animal that will carry it to the stigma of another flower. Shape and colour of the flower often help in attracting pollinators, for example the bright red of many tropical flowers pollinated by small birds, or a star shape (sometimes reinforced by radial lines on the petals, converging towards the reproductive organs of the flower) which is attractive to bees and other insects. Flowers can exhibit colour patterns in the ultraviolet spectrum, invisible to us but very apparent to their insect pollinators. The flowers of many insect-pollinated plants attract pollinators by releasing perfumes – or at least strong scents, including the smell of putrefied flesh emitted by the inflorescences of many plants of the arum family, which is attractive to the flies that act as their pollinators (there is no arguing about taste). Also significant is the widespread presence of floral nectaries, whose secretions are an important source of food for many pollinators, like the pollen itself, which is often produced in quantities greatly in excess of the amount necessary for the reproduction of the plant.

Moreover, many flowers show particular morphological adaptations that make the interactions with their usual pollinators more precise, while excluding other animals. For example, the nectaries of a large flower, placed in the deepest point of a campanulate corolla, can only be reached by an insect with mouthparts shaped like a very long straw. In other flowers, as in those of many orchids, legumes or species of the mint family, a particularly developed petal may serve as an 'airstrip' for the landing of an insect onto which the pollen-filled anthers will release their load. The pollen is often poured onto the visitor by means of a lever mechanism operated by the weight of the insect (as in the sage, *Salvia*) or a spring, whose elastic force is freed by the advancing insect, which displaces the petals that were holding the stamen filaments in a folded position (as in some legumes). Zoophilous pollination requires a fairly precise agreement between the dates and times of day the flowers are open and the dates and times that the potential pollinator is active. This concordance is often evident, as in the case of many flowers pollinated by moths, which open their corollas and/or begin to emit their scent only in the evening; an example is the hoary stock (*Matthiola incana*).

Some plants have mixed pollination systems, relying on structures suitable for both the visit of insects and the release of pollen into the atmosphere. Heather and cyclamen, for example, behave as entomophilous at the beginning of the flowering season, and then as anemophilous later on.

Reproductive Incompatibility

The genetic exchange associated with amphigonic sexual reproduction cannot always occur between any pair of individuals in a population. There are many forms of reproductive incompatibility. The most obvious limitation depends on the fact that an individual belongs to a sex or to a particular mating type which allows it to mate only with individuals of the opposite sex or of a different but compatible mating type. Other forms of reproductive incompatibility are those that prevent *self-fertilization* in hermaphrodite individuals (*self-incompatibility*, or *self-sterility*). In animals, some forms of hermaphroditism (e.g. sequential hermaphroditism; Chapter 4) are naturally secured against self-fertilization.

In haplodiplontic organisms with hermaphrodite sporophytes, including most flowering plants, self-fertilization, strictly speaking, cannot occur, because male and female gametes are produced by two separate gametophytes, even if these are generated by the same sporophyte. However, since the expected genetic similarity between two gametes produced by two gametophytes from the same mother plant is identical to that of two gametes produced by a diplontic hermaphrodite individual, say a land snail, we extend the term self-fertilization to refer to the case where ovules of a flower are fertilized either by pollen from other flowers of the same plant, or by pollen produced by the same flower. Accordingly, in seed plants *cross-fertilization* means fertilization of the ovules by pollen produced by another plant.

In many hermaphrodite plants, self-fertilization is rare or absent. This can be due to several causes. For example, male and female sexual organs of the same individual may mature at different times, or they may occupy positions such as to make self-fertilization unlikely. The barrier to self-fertilization may also depend on alleles that cause self-sterility, most often by preventing development of the pollen tube on flowers of the same plant.

In some species, a mechanical barrier to self-fertilization is provided by the spatial separation of stamens and stigma of the same flower. *Heterostylous* plants produce bisexual flowers of two or three different types, which differ from each other in the relative length of the style and the stamen filaments (Figure 5.1). In those with two types of flowers, there are individuals with 'pin' flowers, where stamens are about half the length of the pistil, next to individuals with 'thrum' flowers, in which stamens are about twice the length of the pistil. In plants with three types of flowers, there is an additional form with a long pistil and stamens of intermediate length. The two (or three) classes of floral morphology are neatly distinct, without

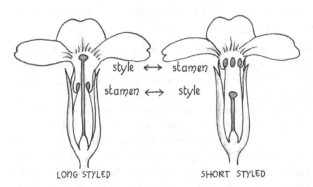

Figure 5.1. Two-morph heterostyly in the primrose (*Primula*). The two morphs differ in style length and stamen height. Pollen from a flower cannot fertilize another flower of the same morph.

intermediates. Each individual plant produces only flowers of the same type, but the two (or three) types are all represented in the population. Pollination leads to fertilization only if it occurs between flowers of two different types, and thus between different plants. The different length ratios between the male and female parts of the flower make it very difficult to exchange pollen between two flowers of the same type, and the transfer of pollen from a thrum flower to a pin flower generally depends on the activity of a pollinator other than the one that could mediate pollination in the opposite direction. Examples of plants with two-type heterostyly are many species of primrose (*Primula*) and flax (*Linum*). The common purple loosestrife (*Lythrum salicaria*) and Bermuda buttercup (*Oxalis pes-caprae*) have three-type heterostyly.

More widespread are the genetic–physiological mechanisms of self-incompatibility, which function in the same way as the mating types in fungi and in many unicellular eukaryotes. In the genome of many flowering plants there are self-sterility genes with very many alleles, sometimes hundreds, which cause the (haploid) pollen that carries a certain allele not to develop a functional pollen tube through the (diploid) tissues of the carpel of a plant that carries a copy of the same allele. In the majority of cases studied thus far, incompatibility depends on a single gene (generally indicated as *S*), but a number of plants have independently evolved a mechanism based on multiple *S* genes. It is estimated that 39% of the flowering plants possess some mechanism of self-incompatibility.

Self-incompatibility can be implemented by either the gametophyte or the sporophyte. In the first case, which is the most common, known for example in many plants of the Papaveraceae, Solanaceae, Rosaceae, Fabaceae, Campanulaceae and Poaceae, rejection of pollen is determined by a factor present in the haploid genome of the male gametophyte and expressed during the growth – soon interrupted – of the pollen tube along its path through the style. In the second case, rejection is determined by the diploid genome of the mother sporophyte; this mechanism, reported for members of the Brassicaceae, Asteraceae, Convolvulaceae, Betulaceae and Caryophyllaceae, is less frequent. The diploid cells that line the inner walls of the anthers, where pollen is produced, deposit on the wall of the latter the specific proteins corresponding to both alleles of the *S* gene. Once the pollen grain reaches the stigma, if incompatible, it fails to hydrate and therefore to germinate and ends up being exhausted without producing the pollen tube.

In an extraordinary case of evolutionary convergence between plants and animals, a self-incompatibility system controlled by two genes has been also described in a sessile animal, the ascidian *Ciona intestinalis*.

Mating Systems

Especially in long-lived animals with a rich social life, such as mammals and birds, the relationship between individuals of the two sexes occurs in a variety of social contexts called *mating systems*. A mating system is the way in which a group of individuals is structured in relation to sexual behaviour: which males and females mate, with how many partners, and under what circumstances. It may also include mate choice, generally by the female, an important component of the evolutionary process of sexual selection (Box 4.1). Two main systems are recognized: *monogamy*, where an individual has only one exclusive partner during a reproductive season, and *polygamy*, where the same individual can have multiple partners during a single reproductive season.

In most cases, monogamy means a pair that lasts for a single breeding season, but it can also lead to the formation of a lifelong stable pair. Examples of seasonal monogamy include brown bears and most passerine birds. Lifetime monogamy is common among birds, much less so in mammals; examples include eagles, swans and gannets among the former, badgers, foxes and jackals among the latter.

Polygamy is a more heterogeneous class, including *polygyny* (a male mating with multiple females), *polyandry* (a female mating with multiple males) and *promiscuity* in the strict sense, in which each individual can mate with any other individual of the opposite sex in the population. Extra-pair matings (occasional episodes of promiscuity in monogamous species) are not uncommon in birds.

Polygyny is the most common polygamous mating system among vertebrates. It leads to a social organization characterized by more or less stable groups, often limited to the breeding season, formed by one male and several females, sometimes called a harem. Polyandry is common among insects and fishes, but is also known in some frogs, turtles and birds. In mammals it is found in the naked mole-rat (*Heterocephalus glaber*), a burrowing eusocial rodent, and has also been recorded in some mustelids, including the European polecat, in cetaceans and in primates (e.g. the marmosets and tamarins of the genus *Callithrix*). Under polyandry, some important processes of sexual selection can be established, such as sperm competition and cryptic female choice. A variant of these promiscuous systems is *polygynandry*, where two or more males have an exclusive relationship with two or more females, forming a stable group where the number of males is not necessarily the same as the number of females. An example is the Bicknell's thrush (*Catharus bicknelli*), a North American passerine.

In some species, an individual can adopt different mating strategies depending on circumstances, which leads to a multiple mating system, as in the case of the dunnock (*Prunella modularis*), a European passerine whose mating system includes monogamy, polyandry, polygyny and polygynandry. Moreover, different individuals of the same species can adopt different mating strategies according to factors such as age, the dominant or subordinate rank they occupy within the group to which they belong, or their specific phenotype (e.g. being relatively large or small, or with developed or underdeveloped sexual characters).

Mating systems have great relevance for the reproductive biology of the species that adopt them and for their evolution, especially in relation to sexual selection. These different interindividual contexts can have important consequences for the parental care that one or the other of the two partners, or both, will later provide to the offspring, an issue discussed in Chapter 8.

A final note on the systems of mating within a population. Contrary to some theoretical expectations, a recent study has revealed little support for the widely held view that animals tend to avoid mating with relatives. These expectations are supported by the same arguments that explain the evolution of the mechanisms of self-incompatibility we have seen in the previous section. Mixing genes with non-relatives (cross-breeding) is considered beneficial because it increases genetic diversity, while mating with relatives (inbreeding) is considered detrimental because of the loss of genetic variation and the possible emergence of genetic defects in the offspring. However, unbiased mating with regards to kinship appears widespread across the animals, so that avoiding mating with relatives is apparently not a must.

Remarkably, mating with very close relatives (*incestuous mating*) is regular or even obligate in some species. We have seen obligate oedipal (mother–son) mating in the mite *Histiostoma murchiei* (Chapter 1), but this also happens in some nematodes (e.g. *Gyrinicola batrachiensis*). Brother–sister mating is obligate in the mite *Adactylidium*. Females carry 6–9 fertilized eggs which hatch when they are still in their mother's body. From the egg clutch a single male develops, which, together with his sisters, feeds on the mother's tissues from inside her body (matriphagy, Chapter 8). Before leaving the mother's remains, the male fertilizes his sisters and dies within a few hours, while his pregnant sisters will look for a thrips egg to feed on, to continue the cycle. Reproduction in the fig wasp *Blastophaga*, which lays eggs inside a still growing fig, where these develop into adults, routinely involves brother–sister mating. For the same reason – the complete development of an egg clutch in a closed environment – brother–sister mating is also frequent among parasitoid wasps.

Summing up, in sexual reproduction there are many ways partners encounter each other (if they do), and multiple options for the gametes of opposite sex to meet and fuse (Table 5.1). However, sexual reproduction does not always take the form of a 'canonical' two-parent affair, as we will see in the next chapter.

Gamete encounters
- external fertilization (gamete fusion in the environment)
  - broadcast spawning (egg and sperm released in the water)
  - external insemination (sperm released in the immediate vicinity of the eggs)
- internal fertilization (gamete fusion within the body of the female)
  - internal insemination (sperm released within the body of the female, through copulation)
  - spermatophores (sperm released in packages outside the body of the female)
  - free spawning (sperm released in the water, eggs fertilized in-situ)

Table 5.1. Summary of the way gametes of opposite sex can meet.

6 One-Parent (or Nearly so) Sexual Reproduction

Self-Fertilization

A zygote does not necessarily derive from the fusion of gametes or gametic nuclei produced by different individuals. Both egg and sperm may instead be produced by the same individual, a sufficient simultaneous hermaphrodite (Chapter 4). In this case, the offspring has only one parent. However, the gametes that merge are the products of independent processes of meiosis undergone by different germ cells, although in the same individual: this distinguishes *self-fertilization* (or *selfing*) from some forms of parthenogenesis where there is the fusion of two of the four nuclei deriving from the same meiosis, as we will see in the next sections (Figure 6.1).

Examples of selfing animals are found among the tapeworms, for example the common pork tapeworm *Taenia solium*. Only one adult specimen of this parasite (up to 5 metres in length) is usually found in the intestine of its host, for example a human. Fertilization may occur between male and female gametes produced by the same proglottid, one of the segments forming most of the tapeworm's body: there are up to 1,000 in a large tapeworm, despite the fact that every day three or four of these detach and end up in the faeces of the host. Fertilization may also occur between gametes produced by two distinct proglottids, but the genetic consequences for the offspring are the same. Selfing is also observed in a few freshwater snails, but usually as a viable option for individuals also able to outcross. A distinct advantage for selfing individuals is that only one of them is required to found a new population. This is a frequent event for pond snails inhabiting very small, temporary water bodies.

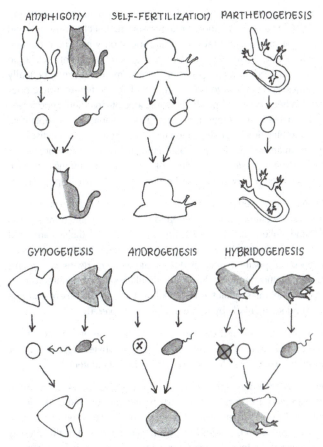

Figure 6.1. Schematic representation of different forms of sexual reproduction. Gray and white colours indicate the different derivations of the genome. See text for a full description.

From a genetic point of view, self-fertilization is comparable to *inbreeding*, the production of offspring from the crossing of individuals that are kin, and thus genetically similar. In a sense, self-fertilization is the most extreme form of inbreeding, since who is more kin than oneself? In principle, in the case of diplontic organisms, the progeny could be genetically diverse and different from the parent, because of the independent assortment of chromosomes and recombination at meiosis. In practice, in a population where reproduction occurs only by self-fertilization generation after generation, genetic variation is progressively eroded, ending up in lines whose members are genetically identical (clones) and homozygous for all genes. This is different from clones resulting from other modes of reproduction (e.g. ameiotic parthenogenesis; see below in this chapter) where intra-individual genetic variation is maintained. However, self-fertilization is usually optional, so that a species can practise both cross-fertilization and self-fertilization, alternating the two modes more or less regularly (*mixed mating system*). The effects of this genetic reshuffling may have sizeable, although temporary, effects on the genetic variation of the population.

Sufficient simultaneous hermaphroditism, in which self-fertilization is possible, is common in some animal groups (ctenophorans, parasitic flatworms such as tapeworms and flukes, pulmonate gastropods, pyrosomid thaliaceans, and a few bony fishes among the Serranidae and Sparidae), but has also been found in a few other species in other phyla (nematodes, nemerteans, bryozoans, phoronids and gnathostomulids). Sufficient simultaneous hermaphroditism is exceptional in vertebrates: in fact, only the killifish *Kryptolebias marmoratus* is known to reproduce regularly by self-fertilization (Chapter 4).

Things are more complex and diverse in plants, because of the alternation between a haploid gametophyte and a diploid sporophyte. Opportunity for selfing must thus be explored for the two generations separately.

Mosses and ferns have hermaphrodite gametophytes, that is, the same green leafy moss plantlet or the same (often heart-shaped) fern prothallus produces both female and male gametes, but these are not the outcome of meiosis, because gametophytes are haploid. As a consequence, self-fertilization is possible, but with the effect of generating a sporophyte that is homozygous for all genes, because the two gametes are genetically identical. However, in most of these plants self-fertilization is not obligate, and the opportunistic resort to different modes of reproduction can be seen as a strategy to cope with different ecological situations, for instance, long-distance dispersal. *Gametophytic selfing* can thus help mosses and ferns establish new populations through wind dispersal of single spores, much more easily airborne than even the smallest seed of seed plants. Generating sporophytes by gametophytic selfing can be a means by which the first migrant of a species to arrive at a distant location persists until further migrants arrive. Cross-fertilization can then increase genetic variability to the level that could have resulted from the far less likely scenario of two or more genetically different spores of the same species landing simultaneously at a suitable distance for gamete encounters (for these organisms, in the order of centimetres).

In plants with separate-sex gametophytes, including all seed plants, there cannot be self-fertilization, because the male and female gametes are actually produced by two distinct individual gametophytes, even if the latter are generated by the same sporophyte. From the point of view of transmission genetics, this *sporophytic selfing* is however equivalent to the self-fertilization of a diploid animal: it consists in the fusion of products of the meiotic divisions (actually, indirect products, because each meiosis has been followed by some mitoses) of two distinct diploid cells of the same sporophytic individual. Of those two meiotic events, one gave rise to the spore that developed into the female gametophyte, the other to the spore from which the male gametophyte developed. In seed plants, self-fertilization (or sporophytic selfing) is actually self-pollination.

Almost all the hermaphrodite flowering plants that regularly practise self-fertilization seem at least occasionally to resort to cross-fertilization. However, in about 70 species (e.g. the four-leaf allseed, *Polycarpon tetraphyllum*, an annual herb of the pink family) self-fertilization is obligate. The flowers of these plants, which are said to be *cleistogamous*, never open, although they reach maturity, a condition mechanically preventing cross-fertilization while allowing self-pollination. Cleistogamous flowers are generally small and inconspicuous, with reduced petals and pollen-poor anthers. In many plants with cleistogamous flowers, however, such as *Viola* and *Impatiens*, flowers that open at maturity coexist with flowers that do not, and therefore cross-fertilization is not completely ruled out.

Parthenogenesis

That virgin female aphids can regularly produce viable offspring was long suspected. In his monumental *Mémoires pour servir à l'histoire des insectes*, René Antoine Ferchault de Réaumur (1683–1757) suggested an experiment to rear aphids in isolation in order to prove this suspicion: he had tried himself several times, but without success. However, Réaumur's book, published in 1737, inspired the young Swiss naturalist Charles Bonnet (1720–1793) to perform his own experiments. In May 1740, Bonnet made his first attempt with a black bean aphid (*Aphis fabae*) kept isolated from the other aphids since the day it was born, and inverted a glass vase over the twig where it sat so that the rim was in close contact with the soil. Checking it twice a day, he was eventually rewarded by the birth of the aphid's first offspring, born at 7.30 p.m. on 1 June 1740, to be followed by another 94 siblings born to this virgin mother over the next 21 days. This experiment won Bonnet the honour of admission as the youngest corresponding member of the Academy of Sciences in Paris. He soon repeated the experiment with other aphid species and eventually set out to rear solitary aphids through up to five generations. Bonnet, however, knew that at least some aphid species had sex, and described males and egg-laying females of the sexual generation of an oak aphid.

In *parthenogenesis* – from the Greek words for generation (*genesis*) by a virgin (*parthenos*) – a new individual develops from an unfertilized egg (Figure 6.1). This form of reproduction occurs through a diverse array of modes, which can lead to different results in terms of genetic variation. Offspring can be only female, only male or both sexes; they can be genetically identical to the mother or different, and either diploid like the mother or haploid. In addition, parthenogenesis can be the only reproductive mode of the species (or population), or alternate more or less regularly with amphigonic reproduction. At the species or population level, but also sometimes for the single individual, parthenogenesis can be accidental, facultative or obligate. And different combinations of these options are possible.

In the cases where the offspring are all females genetically identical to the mother, some authors consider parthenogenesis as a form of asexual reproduction. However, as explained in Chapter 1, here we consider it as a (derived) form of sexual reproduction, because parthenogenesis uses, often in a peculiar way, developmental mechanisms of gametogenesis typical of sexual reproduction, irrespective of its clonal or non-clonal outcome.

Although much less common than amphigonic reproduction, parthenogenesis is a reproductive mode very widespread among multicellular eukaryotes, plants and animals alike. However, the mechanisms and the associated terminology differ quite extensively between the two groups, so we will treat the subject separately in the next two sections.

Parthenogenesis in Animals

In animals, parthenogenesis is known in all major taxa and is very common in a few groups, for instance, in about 40% of the aphid species and about 60% of ostracod crustaceans. Different parthenogenetic lines may have evolved from the same amphigonic ancestor. For example, parthenogenesis has evolved at least five times in the cyprinodontid fish *Poeciliopsis monacha-lucida*. Among the vertebrates, it is striking that no bird or mammal species practises parthenogenesis regularly. As for mammals, it is possible that they cannot abandon amphigonic reproduction because of the phenomenon of *genomic imprinting*. In the course of gametogenesis some genes are chemically marked in such a way that they will function only if transmitted paternally, others only if transmitted maternally. Genomic imprinting is largely, although not exclusively, characteristic of mammals, thought to be associated with the evolution of viviparity and the placental nutrition of embryos.

Based on the genetic/cellular mechanism through which parthenogenesis is implemented, we distinguish between ameiotic and meiotic parthenogenesis, a key difference between them being the potential to produce genetic variation. In *ameiotic parthenogenesis* (or *apomixis*), which is known in rotifers and in various groups of arthropods, eggs are produced without meiosis, so that the offspring are genetically identical to the mother. Whereas, in *meiotic parthenogenesis* (or *automixis*), meiosis occurs, and the diploid condition is restored by duplication of all chromosomes before or after meiosis, or by fusion of two (out of the potentially four) nuclei resulting from this reductional division. Production and conservation of genetic variation depends on the precise cellular mechanism by which it is carried out.

A doubling of the whole diploid set of chromosomes before meiosis is the most common mechanism among the parthenogenetic forms of freshwater planarians (free-living flatworms) and terrestrial oligochaetes, as well as many insects and mites, and the only mechanism occurring in parthenogenetic vertebrates. As with ameiotic parthenogenesis, this form of meiotic parthenogenesis results in *unreduced eggs*, that is, eggs with the same number of sets of chromosome as the germ cells from which they derived. These eggs are thus genetically identical to the somatic cells of the mother and will develop into new individuals identical to her and to each other. Existing genetic variation is maintained, but no new variation is introduced. On the contrary, in most forms of parthenogenesis where the diploid condition is restored through fusion of haploid nuclei deriving from the same meiosis, a certain amount of new variation can be introduced, depending on exactly which nuclei fuse with respect to the lineages of the double division. But this is achieved at the expense of genetic variation in the population: when this form of parthenogenesis is practised regularly through many generations, in the long term it tends to erase genetic variation. Species that reproduce by parthenogenesis through the fusion of nuclei deriving from the same meiosis are found among the nematode worms, some freshwater oligochaetes, isopod crustaceans and tardigrades (microscopic animals known as water bears), as well as in various insect groups.

Rather than looking in detail at the nuclear events and their consequences for genetic variation, one can focus instead on the sex of the individuals generated. Parthenogenesis that generates only females (*thelytokous parthenogenesis*) is the most widespread form in animals. It is the parthenogenesis of the summer generations of species with alternation of amphigonic and parthenogenetic generations (Chapter 2), such as water fleas, rotifers, aphids and other insects. Parthenogenesis that produces both males and females (*amphitokous* or *deuterotokous parthenogenesis*) is instead typical of the late-summer generations of these animals, where the production of the first males is a prerequisite for a return to the amphigonic reproduction that will close the seasonal cycle.

Parthenogenesis that generates only males (*arrhenotokous parthenogenesis*) is accompanied by other forms of reproduction: a population would not survive if made up of males only. Outside animals, in organisms with moderate levels of gamete dimorphism, such as some brown algae, development from male (but also female) gametes in the absence of fertilization occurs frequently, at least under laboratory conditions. However, in the more common case of a marked gamete dimorphism, as in animals, arrhenotokous parthenogenesis is associated with a peculiar system of sex determination, the haplodiploid system. This is based on a different chromosomal asset in the two sexes: females are diploid, males haploid (Chapter 7). This requires the coexistence of parthenogenesis and amphigony in the population: unfertilized eggs develop into males, fertilized eggs develop into females. This kind of male-only parthenogenesis is limited to a few animal taxa, and is unknown in plants. In hymenopterans (bees, wasps and ants), it is the rule, with few exceptions, and lies at the basis of the social organization in these insects.

For instance, in many ant societies, winged males emerge from pupae along with the usually winged breeding females (prospective queens) of the same generation. Winged females and winged males leave the colony, disperse and eventually mate, in what is called a nuptial flight. This usually takes place in the late spring or early summer when the weather is hot and humid. Heat makes flying easier and freshly fallen rain makes the ground softer for mated queens to dig nests. Females of some species mate with just one male, but in others they may mate with as many as 10 or more different males. Sperm are stored in the queen's abdomen in a special organ known as a spermatheca, where they last in a viable state throughout her lifetime. This can be as long as 20 years, during which time the sperm can be used to fertilize tens of millions of eggs. Inseminated females seek a suitable place to establish a new colony. There, they break off their wings using their tibial spurs and begin to lay and care for eggs. The females can selectively fertilize eggs with the sperm stored to produce diploid females (either queens or workers) or lay unfertilized haploid eggs to produce males. Workers cannot mate; however, workers of a number of species lay unfertilized eggs that become fully fertile, haploid males. These workers are thus obligate parthenogens.

Parthenogenesis in Plants

In botanical terminology, processes equivalent to those described in animals as parthenogenesis are commonly referred to as *apomixis*, while in the zoological tradition this term indicates ameiotic parthenogenesis, one of the two main mechanisms of parthenogenesis. Anyway, the fundamental point is that in parthenogenesis, of plants and animals alike, the offspring develops from an egg that does need to be fertilized.

Parthenogenesis is very widespread in ferns, but accidental in gymnosperms, no species of which routinely adopts this reproductive mode.

In the case of angiosperms, parthenogenesis is often described as a form of vegetative (asexual) reproduction through seed. However, beyond the choice of words, plant parthenogenesis should not be confused with asexual reproduction. The main distinctive features of the two modes of reproduction are that vegetative reproduction does not necessarily pass through a single-cell stage: most propagules are multicellular, unlike the unfertilized egg, and daughter plants produced by vegetative propagation generally settle close to the mother plant, while seeds from pathenogenesis are very often dispersed over greater distances. Moreover, seeds can enter a phase of dormancy, thus resisting adverse seasons or conditions, a strategy that is not available to vegetative propagules.

As in animals, the mechanism through which a parthenogenetic egg can be obtained are very diverse, with different effects on the maintenance and production of genetic variation, but botanists distinguish two main pathways. In *gametophytic apomixis*, the female gametophyte is conserved, but the female spore is diploid and so is the female gametophyte that develops from it. This form of uniparental reproduction is found in *Hieracium*, *Taraxacum* and *Allium*. Depending on the mechanisms involved, which may pass through meiosis or skip it, this form of apomixis can generate offspring genetically identical to the mother plant or not. Apomictic clones are very common among the dandelions (*Taraxacum*) that live in disturbed environments, where the simultaneous flowering of a huge number of plants reveals at a glance that they belong to the same clone. However, sexual reproduction in *Taraxacum* is more common than generally assumed, and even characterizes local minorities of plants in mixed populations.

In *sporophytic apomixis*, the other main type of plant parthenogenesis, embryos are formed by diploid cells of the ovule that surround the female gametophyte, and thus actually belong to the sporophyte generation. These cells do not give rise to a gametophytic generation, which is skipped. The life cycle is reduced to monogenerational. This is observed in many orchids and in the genus *Citrus*, which includes oranges and lemons.

To make the process even more peculiar, in most parthenogenetic species of flowering plants, the endosperm, the nutritive tissue of the seed, develops only after fertilization by *pseudogamy*, a modification of the normal double fertilization (Chapter 5) that requires pollination, although the egg does not need to be fertilized. In many ways this recalls gynogenesis in animals (see the last section of this chapter). Failure to develop endosperm leads to seed abortion. Therefore, parthenogenetic plants generally maintain the male function, acting as pollen donors in crossings with amphigonic lines or lines with non-obligate parthenogenesis. However, some parthenogenetic species of the sunflower family have developed a form of apomixis independent of pollen, in which endosperm can develop without being fertilized.

Natural History of Parthenogenesis

According to traditional descriptions, some plant and animal species include both amphigonic and parthenogenetic populations, the latter often polyploid (i.e. with more than two sets of chromosomes per nucleus). Such an account, however, is at odds with the so-called *biological species concept*, possibly the most popular among the very many species concepts formulated by biologists thus far. This defines a species as a set of actually or potentially interbreeding populations which are reproductively isolated from other populations. However, it is logically impossible to apply the biological species concept to organisms with uniparental reproduction (no breeding, no interbreeding), so it is perhaps better to say that some amphigonic species are accompanied by exclusively parthenogenetic populations derived from them.

The most relevant aspect of this phenomenon is the geographical and ecological distribution of populations with different reproductive modes, and this is why parthenogenesis in this context is called *geographical parthenogenesis*. Parthenogenetic populations usually occupy marginal areas of the species' range, subject to difficult or even extreme environmental conditions, and are unusually quick to colonize new areas. A typical example can be seen in the almost always polyploid parthenogenetic populations of weevils (Curculionidae) in Alpine areas and in the northernmost regions of Europe, which have been free of ice for only a few thousand years, or even less. In the psychid moth *Dahlica triquetrella* multiple transitions to parthenogenesis occurred in different lineages during the last ice age. The greater capacity for colonization demonstrated by these populations can be partly attributed to the short-term advantage of uniparental reproduction (i.e. a single individual can found a new population); in part, however, it seems to be due to their polyploid condition. In plants as well, patterns of geographical parthenogenesis are probably caused by a combination of factors that would explain why parthenogenetic lineages often have larger distributional ranges or reach higher latitudes and altitudes than their amphigonic relatives.

Accidental parthenogenesis is observed in many species which are normally amphigonic, for example in some birds such as zebra finch, chicken, pigeon and especially turkey, where the phenomenon can affect one egg in five. Parthenogenesis is also known in the Indian rock python, the Komodo dragon (Chapter 2), the bonnethead shark, some species of *Drosophila*, stick insects and many mayfly species. It is likely that some obligate parthenogenetic species evolved from species with accidental parthenogenesis as a first step, but the persistence of accidental parthenogenesis in otherwise amphigonic species can also be viewed as an optional reproductive strategy, when a few founders colonize a new territory, when small populations are distributed at the edge of a species' range, or, as in the case of the mayfly, when low dispersal ability and short adult life can limit mating opportunities.

Reproducing while still in a juvenile stage is called *paedogenesis*. In the most common cases, paedogenesis is achieved through optional *larval parthenogenesis* or, in insects with an intermediate pupal stage between the larva and the adult (holometabolous insects), through *pupal parthenogenesis*. In insects, paedogenesis has evolved at least six times independently. The phenomenon is particularly widespread in the gall midges (Cecidomyiidae), several of which reproduce in the larval (*Miastor*, *Heteropeza*; Chapter 2) or pupal (*Tekomyia*, *Henria*) stage. In these insects, parthenogenesis is a reproductive option, an alternative to amphigony. However, in the unique paedogenetic beetle *Micromalthus debilis* the adults are apparently no longer reproductively functional: amphigonic reproduction has been lost as an option, and despite the presence of rare 'ghost adults', the species actually reproduces exclusively by larval parthenogenesis.

In at least 15,000 species belonging to several animal clades, parthenogenesis alternates with amphigony in a heterogonic life cycle (Chapter 2). This *cyclical parthenogenesis* occurs in most representatives of the monogonont rotifers, digeneans, cladoceran crustaceans (water fleas) and cynipid hymenopterans, as well as in many aphids. In monogonont rotifers, the males are haploid, while the females are diploid (as in hymenopterans; Chapter 7) and can produce two types of egg. During the favourable season, females produce large diploid eggs that do not need to be fertilized and develop into females within 12–48 hours. After several parthenogenetic generations, typically in autumn, females develop with a slightly modified ovary, compared to that of the previous generations. These females produce haploid eggs, regularly obtained by meiosis. These, if fertilized, produce a thick shell, become resistant and develop into mature (diploid) females the following season. If these haploid eggs are not fertilized, in the same season they develop into males, which can mate with their 'aunts'. Parthenogenesis allows rapid growth of the population, while amphigonic reproduction enable the rotifers to survive through the adverse season or shorter unfavourable periods.

In interspecific hybrids, in which meiosis is generally difficult or impossible because chromosomes cannot pair up before dividing, natural selection favours cellular mechanisms that allow the production of diploid eggs able to develop anyway, either following fertilization (which however leads to triploidy, an inconvenient (odd) chromosomal complement in itself) or, better, without fertilization, that is, by parthenogenesis. This *parthenogenesis of hybrid origin* is probably common to all uniparentally reproducing vertebrates. Among the lizards there are about 30 parthenogenetic species. Further examples are found in gastropods, crustaceans, curculionid beetles, stick insects and orthopterans.

Parthenogenesis is sometimes *induced by parasites*. Bacteria of the genus *Wolbachia* are endocellular parasites of numerous arthropods and nematodes, in which they can be transmitted from mother to offspring through the cytoplasm of the egg. The bacterium has evolved strategies that allow it to increase its diffusion by manipulating the sex ratio of the host in favour of females. For example, they induce thelytokous parthenogenesis in species with a haplodiploid sex-determination system (bees, ants, wasps and certain mites; Chapter 7). Unfertilized eggs of these arthropods, otherwise expected to develop into males, undergo a genome duplication triggered by the parasite, after which, now diploid, they develop into females. Parthenogenesis is similarly induced in their hosts by some microbial parasites other than *Wolbachia*.

Other Forms of Sexual Reproduction

The fish *Poecilia formosa* is a natural hybrid between *P. mexicana* and *P. latipinna*. This all-female species produces unreduced, diploid eggs. However, these cannot develop into adults by themselves, as in parthenogenesis, but need to be 'stimulated' by the sperm produced by males of a related, amphigonic species of *Poecilia* occurring in the same waters. This is an example of those unconventional modes of sexual reproduction that involve eggs and sperm cells, but deviate from amphigony because the genomes of the two gametes do not merge in the zygotic genome.

The reproductive mode employed by *P. formosa* is known as *gynogenesis*, where the male gamete comes into contact with the egg cell, or penetrates it, but does not contribute any genetic material to the genome of the organism that develops from it (Figure 6.1). In other words, the egg needs to be activated by the sperm, but it is not fertilized by it. A gynogenic species such as *P. formosa* is not hermaphrodite; it is composed of females only, and the stimulus to the development of the eggs is provided by (non-fertilizing) insemination by males of closely related species. Gynogenesis is known for scattered species of several groups, including freshwater planarians, insects, fishes and amphibians. Salamanders of the genus *Ambystoma* are classic examples of gynogenic organisms, and are believed to have been reproducing this way for over a million years. It is probable, however, that in these animals the fertilization of an egg, capable of introducing new alleles into the gene pool, is a rare but not impossible event.

Androgenesis is in effect the opposite of gynogenesis (Figure 6.1). Following the penetration of the sperm cell into the egg, only one of the two genomes is retained in the zygote, but in this case it is the paternal one, because the chromosomes of the egg are absent or inactivated. For instance, in some stick insects of the genus *Bacillus* two male sperm nuclei can meet and merge inside an egg in which the maternal genome has degenerated or has been lost. This can produce offspring of both sexes, while from the point of view of genetic variation the effect will be similar to self-fertilization. Things are different when the genome of an unreduced, diploid sperm cell replaces the haploid genome of the egg cell, as in four species of the freshwater clam *Corbicula* and in the freshwater fish *Squalius alburnoides*, the first-known vertebrate to exhibit androgenesis. In these cases, the nuclear genome of the offspring will be an identical copy of the genome of the sperm-donor parent, and if the species is not hermaphrodite and sex determination is genetic, the line will be potentially all-male. The androgenetic Saharan cypress (*Cupressus dupreziana*) produces diploid pollen (male gametophyte), whose sperm nuclei can penetrate the eggs of the same species, but also those of another species, *Cupressus sempervirens*, in both instances entirely replacing the egg genome with the genome of the pollen grain.

A notable case is provided by the little fire ant (*Wasmannia auropunctata*), where haploid males are generated by androgenesis, while the queens generate other queens by parthenogenesis. Normal amphigonic reproduction brings male and female genomes together only in the workers, but these are sterile. This creates a singular situation for a sexually reproducing species, where males and females reproduce without exchanging genetic material, as if they were two distinct species. Recognizing them as such would result in a unique case of a species composed exclusively of males.

Another unconventional form of sexual reproduction is *hybridogenesis*, a reproductive mechanism midway between biparental and uniparental sexual reproduction. In its simplest form, a hybridogenic female develops from fertilized eggs, where maternal and paternal genomes originally coming from two different species coexist and are regularly expressed. However, only the maternal genome is transmitted to the next generation in the eggs of these females, so that they have to mate with males of the species corresponding to the genome expunged from the eggs (Figure 6.1). These males are the genetic fathers of their children, but cannot be the genetic grandfathers of their grandchildren.

One of the most thoroughly investigated hybridogenic systems is found in the European green frogs currently assigned to the genus *Pelophylax*. The overall picture is very complex and involves several species, so we present here only the most frequent condition in the system traditionally described as a complex of three species, *P. lessonae*, *P. ridibundus* and *P. esculentus*. The latter is actually a hybrid between the other two species and is the prototype of reproduction by hybridogenesis. In the hybrid, gametogenesis involves the elimination of the *P. lessonae* genome, so that the females of *P. esculentus* can only produce eggs with the *P. ridibundus* genome. The continuity of the hybrid through the generations is therefore dependent on the availability of male individuals of the other parental species, *P. lessonae*. Consequently, *P. esculentus* can survive only under local coexistence with the latter. Although less frequent, there are also hybrid populations that transmit only the *P. lessonae* genome and therefore can only reproduce with the contribution of genomes provided by males of *P. ridibundus*.

Summing up, in this chapter we have seen many ways in which a single parent can reproduce sexually by itself, or nearly so (Table 6.1). However, in order to reproduce sexually, either alone or as one of a pair, organisms must develop reproductive competence. How they do this is something we will explore in the next chapter.

Uniparental sexual reproduction
- self-fertilization: egg and sperm produced by the same hermaphrodite individual
- parthenogenesis: reproduction through eggs that do not need to be fertilized
- gynogenesis: reproduction through eggs that do not need to be fertilized, but need to be activated by sperm
- androgenesis: reproduction through eggs where the maternal genome has been replaced by the paternal one, carried by the sperm
- hybridogenesis: reproduction where maternal and paternal genomes of an individual are from two different species, but only the maternal genome is found in the eggs, so that females have to mate with males of other species

Table 6.1.   Summary of different modes of sexual reproduction from a single parent.

7 Development of Sexual Traits

Sex Determination: A Complex Affair

Acquiring the traits specific to a given sex, during early development or at another point during the life of an organism, is usually a complex process. Although the sex condition of an individual is conventionally defined based on the type of gametes it is able to produce (Chapter 4), the sex-specific phenotype is generally not limited to the organs of reproduction. Each of these characters can maintain a certain degree of independence from other sexual traits in the same organism, be subject to different developmental control, and show different degrees of sensitivity to the environment. Therefore, sexual differentiation extends to the development of the secondary sexual characters, which can be morphological, physiological, behavioural, or combinations of these. An exploration of this fascinating subject requires some preliminary clarification about systems and mechanisms of sex determination and sex differentiation.

Despite the greater complexity of sexual differentiation, in most separate-sex species the male-type or female-type expression of the different sexual characters tends to be highly coordinated. Thus, even in cases where the development of sexual characteristics extends over a long time in development, it is usually possible to identify a key factor that is responsible for the generally early 'developmental decision' to take one or the other of the two alternative developmental options, male versus female. For this reason, *sex-determination systems* are traditionally classified based on the nature of the primary causative agent in the specification of an individual's sex.

A main distinction is between *genetic sex determination* and *environmental sex determination*. In the first case, sex is established early in development by genetic factors such as the presence of certain chromosomes, genes or alleles. In contrast, in environmental sex determination the sex of an individual is established by environmental signals, such as temperature values, which are received and interpreted by the individual during its development. A third category, which borrows something from each of the other two, is *maternal sex determination*, in which the sex of the offspring is determined either by the genotype or by a physiological condition of the mother. Finally, many organisms have *mixed sex-determination systems*, where genetic and environmental factors are combined to different degrees.

These sex-determining signals, whether genetic or environmental, can be associated with a number of developmental *mechanisms of sex determination* that interpret them. For instance, the very common XY chromosomal sex-determination system (XY male, XX female) can be associated with a sex-determining mechanism where the sex of an individual is decided by the presence of the Y chromosome (as in mammals), or alternatively by a mechanism sensitive to the number of X chromosomes (as in the fruit fly *Drosophila*).

Mechanisms of sex determination gradually blend into the processes of *sexual differentiation*, crossing the traditional but arbitrary boundary between biology of reproduction and biology of development. This particular aspect of development, although obviously related to the primary determination of sex, is to some extent independent of it.

The sex-determination system, the mechanism that implements it and the sexual differentiation that follows may differ greatly among closely related species, and even within one species. For instance, intraspecific variation in the sex-determination system has been described for the house fly (*Musca domestica*) and a small rodent, the lemming *Myopus schisticolor*.

Sex determination does not apply to simultaneous hermaphrodites, where an individual can produce eggs and sperm at the same time, but it does apply to sequential hermaphrodites, in the very peculiar form of a switch from one sex function to the other. Moreover, in species with a mixed breeding system, which have both unisexual (male and/or female) and hermaphrodite individuals (Chapter 4), rather than sex-determination systems, we find systems of determination of the sex condition: male versus hermaphrodite, or female versus hermaphrodite, or male versus female versus hermaphrodite. To a large extent, these systems are based on the same principles as sex determination with male versus female options. In much of the following, unless otherwise specified, sex determination is intended in a broad sense, as determination of the sexual condition, which in some cases may also apply to hermaphrodites.

Common Genetic Sex-Determination Systems

In *genetic sex determination*, a gene, a complex of genes or the entire chromosomal complement is responsible for initiating a cascade of developmental events that produce the phenotypic characteristics of either sex. Very often, genes involved in the control of this developmental option are located on a single pair of homologous chromosomes, called the *sex chromosomes*, which occur in two distinct versions, characterized by a different set of genes or by different alleles of the same genes. An example is seen in our X and Y chromosomes.

Sex chromosomes are found in most animals and in a number of plants, where they have evolved many times independently. For instance, in flowering plants, where the separate-sex condition is found in the sporophyte of only 6% of species (about 18,000 species), different genetic systems of sex determination have evolved in more than 100 lineages – and in comparison with animals this has occurred quite recently.

Most common among the genetic systems are *chromosomal sex-determination systems* (Figure 7.1). In these systems, one or more sex chromosomes are present in unequal combination in the two sexes, unlike the other chromosomes (*autosomes*). The two main systems of chromosomal sex-determination for diploids are the XY system, where males have two different sex chromosomes (X and Y) and females have two copies of one of the two (XX), and the ZW system, where females have two different sex chromosomes (Z and W) and males have two copies of one of the two (ZZ). Derived from these two systems, and also common, are X0 (X-zero), where there is no Y and just one X in the male, and Z0 (Z-zero), with just one Z and no W in the female. The sex presenting different (e.g. XY) or unbalanced (e.g. X0) sex chromosomes is called

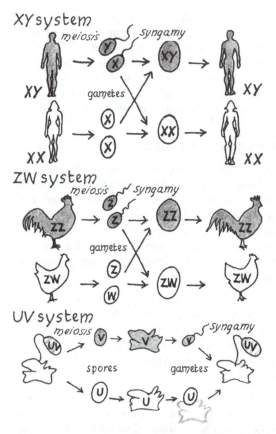

Figure 7.1. Schematic representation of the main chromosomal sex-determination systems: XY, ZW and UV. Grey, males; white, females; striped, sexually indeterminate diploid phase of the UV system. In both the XY and ZW systems, sex determination occurs in the diploid phase, in the UV system in the haploid phase. The heterogametic sex is the male in the XY system, the female in the ZW system.

the heterogametic sex, because it produces different types of gametes (e.g. in the XY system, one half with the Y chromosome, one half with X), while the gametes produced by the other (homogametic) sex all have the same heterochromosome (an X, in the example).

The UV system is instead the most common chromosomal sex-determination system in the haploid phase, as in some haplontic algae and in the haploid generation (gametophyte) of bryophytes. Here, females are characterized by the possession of a single sex chromosome U, while males have a single V.

In both animals and plants, female heterogamety (e.g. ZW and Z0) is less common than male heterogamety (e.g. XY and X0). However, female hetero-gamety is characteristic of two large groups, birds and lepidopterans (butter-flies and moths). It is also found in some crustaceans, fishes, amphibians and snakes. In domestic stocks of the aquarium fish *Poecilia sphenops*, both XY and ZW systems occur. In land plants female heterogamety is found in 10% of the species with genetic sex determination that have been studied. An example is the ginkgo.

As mentioned above, the chromosomal asset of an individual is translated into a developmental decision toward one sex or the other through very diverse genetic mechanisms. These can be grouped into two main categories, those with a dominant sex chromosome and those with genetic balance.

There are sex-determining mechanisms where the sex chromosome that is found exclusively in the heterogametic sex (Y or W) carries genes specifically involved in sex determination (these are known as mechanisms with a *dominant sex chromosome*). In mammals, for instance, the Y chromosome carries a gene not occurring on the X chromosome, the *sex-determining region Y* (*SRY*), which induces the development of the gonads as testes rather than ovaries. Thus, individuals with an abnormal chromosome complement XXY (one extra X) are male, while X0 individuals (Y absent) are female. In contrast, there are also mechanisms where the sex of an individual is settled in the absence of specific determinants for the heterogametic sex on the chromo-somes Y or W (these are known as mechanisms with *genetic balance*). For instance, in *Drosophila*, which shares with mammals an XY system, sex is correlated to the ratio (X:A) between the number of X chromosomes and the number of autosome sets (normally two). Individuals with a ratio X:A $\geq$ 1 are female and those with X:A $\leq$ 0.5 are male, while individuals with an X:A ratio between 0.5 and 1 (e.g. XX associated with a triploid complement of autosomes, X:A = 2/3) are sterile and present intermediate male and female features (intersex individuals). Contrary to the case of mammals, therefore, fruit flies with an unbalanced XXY chromosomal complement (one extra X, X:A = 1) are female, while individuals with X0 karyotype (Y absent, X:A = 0.5) are male. The genes on the *Drosophila* Y chromosome therefore have no effect on the determination of sex, although they are necessary for sperm differentiation and therefore for male fertility. In both animals and plants, the two types of mechanism are well represented.

Finally, an example of the extreme evolutionary lability of the sex-determination systems: in the Japanese frog *Glandirana rugosa*, different populations have different chromosomal sex-determination systems, either ZW or XY.

Less Common Genetic Sex-Determination Systems

A variety of chromosomal sex-determination systems deviate significantly from the more common XY and ZW systems, entailing an extraordinary range of associated mechanisms of sex determination. Here we will limit ourselves to an excursion into the variety of systems.

Some chromosomal sex-determination systems involve more than two sex chromosomes. By analogy with the XY and ZW systems, these *systems with multiple sex chromosomes* are named using formulae that indicate the set of sex chromosomes found in the heterogametic sex, for instance, X_1X_2Y or XY_1Y_2 (corresponding to homogametic karyotypes with twice as many Xs and no Ys in the other sex, i.e. $X_1X_1X_2X_2$ and XX, respectively). Examples among the mammals are the rat-kangaroo *Potorous tridactylus*, the wallaby *Wallabia bicolor*, the shrew *Sorex araneus*, the gerbil *Gerbillus gerbillus* and several leaf-nosed bats, all with an XY_1Y_2 system, while the mouse *Mus minuteides*, the barking deer *Muntiacus muntjak* and several species of mongoose of the genus *Herpestes* have an X_1X_2Y system. Most spiders (85% of the cases studied) have multiple sex chromosomes, typically $X_1X_2$0, but also $X_1X_2X_3$0. No species of spider has Y chromosomes.

Among the most extreme examples of multiple sex chromosomes, the case of the darkling beetle *Blaps polychresta* stands out, with a system involving 12 X and 6 Y chromosomes, that is, $X_1X_2X_3X_4X_5X_6X_7X_8X_9X_{10}X_{11}X_{12}Y_1Y_2Y_3Y_4Y_5Y_6$. The most complex systems in mammals are found among the monotremes, with $X_1X_2X_3X_4X_5Y_1Y_2Y_3Y_4Y_5$ in the platypus. Multiple sex-chromosome systems are found also in animals with female heterogamety: in the snake *Bungarus caeruleus*, females are Z_1Z_2W and males are $Z_1Z_1Z_2Z_2$, while the copepod crustacean *Diaptomus castor* has a $Z_1Z_2Z_3W_1W_2W_3$ system. Very unusual is the New Zealand frog *Leiopelma hochstetteri*, where females are 0W and males 00 (no sex chromosomes).

Also unusual, but for a different reason, are the sex-determination systems of some rodents. Many species of voles of the genus *Ellobius* have XX females and XX males, while in *Ellobius lutescens* and in the rat *Tokudaia osimensis* both sexes are X0. In another vole, *Microtus oregoni*, females are X0, males XY. Two species of lemming, *Myopus schisticolor* and *Dicrostonyx torquatus*, have XX and XY females and XY males. Even stranger, the poeciliid fish *Xiphophorus maculatus* has a peculiar chromosomal system with three sex chromosomes, X, Y and W: males are XY or YY, females XX, WX or WY. It has been hypothesized that this fish is in a phase of evolutionary transition from an XY to a ZW system.

Systems with multiple sex chromosomes are also found in plants: in some conifers of the genus *Podocarpus* (X_1X_2Y), in a variety of hop (*Humulus lupulus* var. *cordifolius*, $X_1X_2Y_1Y_2$), in sorrel (*Rumex acetosa*, XY_1Y_2) and in the mistletoe *Viscum fischeri* (female $X_1X_2X_3X_4$, male $Y_1Y_2Y_3Y_4Y_5$). The UV system has also occasionally evolved into a multiple sex-chromosome system, as in the liverwort *Frullania dilatata*, where the female gametophyte has a U_1U_2 karyotype while the male has V.

In some genetic sex-determination systems there are no sex chromosomes, but males and females have distinct alleles at specific loci on the autosomes. In recent years there have been important advances in the study of these systems, thanks to the application of modern genomics techniques that allow the identification and characterization of genome regions involved in sex determination even in the absence of sex chromosomes. These are not necessarily limited to (or dominated by) a single locus. For instance, in a small freshwater fish, the separate-sex poeciliid *Xiphophorus helleri*, which does not possess sex chromosomes, sex is genetically determined by a set of factors distributed on several chromosomes with masculinizing or feminizing effects, none of which is prevailing, and whose collective balance translates into the development of either sex.

In some animal groups there are no sex chromosomes, but females develop from fertilized eggs and are thus diploid, while males are haploid, developing from unfertilized eggs (reproduction by parthenogenesis; Chapter 6). The apparent simplicity of this *haplodiploid sex-determination system*, typical of hymenopterans, actually hides a difficulty at the level of the genetic mechanisms that implement it. How is it possible that two copies of the same genome determine development of the embryo into a female, while a single copy of *the same* genome determines the development into a male? No gene present in one sex is lacking in the other, and the ratio between the genes involved in sex determination and the other genes is the same in both diploid and haploid conditions. In hymenopterans, diploid individuals that are heterozygous at a locus with very many alternative alleles, called *complementary sex determiner* (*csd*), develop as females, whereas individuals with a single allele, either because haploid or, more rarely, homozygote resulting from inbreeding, develop as males. This sex-determination mechanism has been confirmed for more than 60 species of hymenopterans. In the honey bee (*Apis mellifera*), the *csd* gene has no fewer than 15 alternative alleles. Unfortunately, little is known about the genetic basis of haplodiploid sex determination in animals other than hymenopterans. No plant is known to have this sex-determination system.

Finally, in genetic sex-determination systems, the sex of an individual is not necessarily established by the genome resulting from fertilization (or the lack of it). In some animals the sex-specific karyotype of an individual depends on modifications of its original chromosome complement during an early stage of embryonic development. For instance, in the fungus gnat *Sciara* all zygotes have the same genotype, with three X chromosomes (XXX). The loss of one or two X chromosomes determines whether the zygote will develop into a female (XX) or a male (X0). In many species of scale insects, with a haplodiploid system, elimination or inactivation of the whole set of paternal chromosomes during early embryonic cleavage results in a functionally haploid embryo that will develop into a male. On the other hand, if the diploid condition is maintained during development, the embryo will develop into a female. This genetic mechanism, based on paternal genome loss, is also known for some other taxa, including some mites.

Environmental Sex-Determination Systems

In many species, the sex of an individual is not established by the genetic make-up of its founding cell (e.g. the zygote, or the spore), but is defined through one or more stages of its embryonic or post-embryonic development, depending on factors external to the organism. These can be the temperature to which it is subject during development, or signals coming from other individuals of the same species. In many cases of this *environmental sex determination*, sex is established irreversibly during a restricted, usually early, temporal window, during which the organism is sensitive to a specific external signal that can elicit the developmental path for either sex. In other cases, the environmental determination of sex is associated with forms of sequential hermaphroditism, so that the sex of an individual changes, sometimes more than once, in the course of its life.

Until recently, in developmental biology, environmental effects have been greatly underestimated, acknowledged to have only a permissive role: suitable environmental conditions allow normal development, unsuitable conditions forbid it. This is partly due to the fact that most of what we know about development comes from studies on so-called 'model organisms'. This is a small set of intensively studied species that because of their characteristics (e.g. easy to rear, short generation time) are suitable subjects for laboratory experiments. The most popular among them are the fruit fly *Drosophila melanogaster*, the nematode worm *Caenorhabditis elegans*, the mouse *Mus musculus*, the cress *Arabidopsis thaliana*, the green alga *Chlamydomonas reinhardtii*, the yeast *Saccharomyces cerevisiae* and the bacterium *Escherichia coli*. However, among the characteristics that make them so convenient for research there is also the ability to develop in the laboratory in a highly repeatable way, that is, in a way that is as independent as possible from external factors. Therefore, in this regard, model species are scarcely informative when it comes to environmental effects on development.

Nonetheless, the environment can have an instructive role in development, and in recent times this role in the normal development of various organisms has been increasingly appreciated. Environmental sex determination is a form of a more general biological phenomenon called *phenotypic plasticity*, where individuals with the same genotype can develop different phenotypes as a response to differences in their environment. This is a very active area of research in these times of climate change.

Among the many factors able to induce sexual differentiation in animals or plants are day length, water acidity, nutrition and parasites. In the case of animals, the most common factors are temperature and interactions with other members of the same species.

Unexpected perhaps, but there are examples of *temperature-dependent sex determination* even among the vertebrates. In many reptiles, sex is irreversibly established by the temperature to which the embryo is exposed during incubation (Figure 7.2). In most turtles, all eggs incubated at high temperatures develop into females, while those incubated at low temperatures develop into males. The opposite is observed in some lizards. In other turtles and in crocodiles, incubation at intermediate temperatures will result in males only, while high and low temperatures will result in females. In all cases, the temperature ranges within which both males and females can develop are quite narrow. Temperature-dependent sex determination has been described for several species of fishes and insects as well. Global warming represents a serious threat for species with temperature-dependent sex determination, since it can bias the numerical ratio between sexes in populations and the expression of certain sex traits.

Spatial proximity to other members of its own species, or the way in which an individual interacts with conspecifics, is sometimes responsible for determining its sex, reversibly or irreversibly. In *social sex-determination systems*, the signal that evokes the developmental response for one sex or the other is often chemical (i.e. pheromones), or based on other communication channels, for example tactile or visual. The sex of the marine annelid *Bonellia viridis* is irreversibly established during early development, depending on where the planktonic larva settles to start metamorphosis to the sedentary adult. Larvae that settle on a sea-floor area far from other individuals of the same species

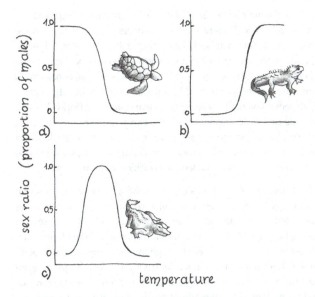

Figure 7.2.   Temperature-dependent sex determination: schematic representation of the relationship between sex ratio and incubation temperature in reptiles. (a) In the loggerhead turtle (*Caretta caretta*), males develop at low temperatures, females at high temperatures. (b) In the tuatara (*Sphenodon punctatus*), females develop at low temperatures, males at high temperatures. (c) In the American alligator (*Alligator mississippiensis*), females develop both at low and high temperatures, males at intermediate temperatures.

develop into females. In contrast, if a larva contacts an adult female, it begins to develop into a male through the effect of a masculinizing pheromone produced by the female. After a few days the male, which remains small (1–3 millimetres) compared to the female (up to 1 m), and in comparison to the latter also has a very simplified body organization, will enter the body of the female, where he will no longer lead an independent life, but will find nourishment and only have to provide for the fertilization of his partner's eggs. From a functional point of view, the male of *Bonellia* is actually reduced to a symbiotic provider of sperm.

There is social sex determination also in the marine gastropod *Crepidula fornicata*, but in this case sex differentiation is not necessarily definitive. In the transition from the planktonic to the sedentary phase of their life cycle, *Crepidula* larvae settle preferentially on the shell of an already settled conspecific, forming stacks of individuals of different ages. For a while, recently metamorphosed individuals are male, but this is followed by a period of sexual lability, accompanied by the degeneration of the male reproductive system. The individual can then become male or female, depending on the composition of the stack. If it fixes on a female, it will become a male; if it detaches from the stack, it will become a female. In the presence of numerous males in the stack, some of these may become females. In any case, the female condition is not reversible.

Many sequential hermaphrodite fishes can also change sex based on social interactions (Chapter 4). In the goby *Trimma okinawae* and a few other species, social interactions can induce an individual to change sex several times (*alternating hermaphroditism*). Changes in the composition of the social group, perceived by the nervous system through the sense organs, can modify the levels of sex hormones of an individual within a few hours or even minutes. The subsequent visible changes affect morphology and behaviour alike. Looking at the temporal schedule of the process of sexual transformation, changes in behaviour generally precede those in the gonads.

Sex determination based on interactions with conspecifics can be found also in plants. In the fern *Ceratopteris richardii*, a spore can develop either into a male or into a hermaphrodite gametophyte (prothallus). Hermaphrodite gametophytes develop from spores in the absence of signals from conspecific gametophytes. These secrete a pheromone that induces the development of male gametophytes from the spores exposed to its action. However, the male sex condition of the gametophyte is not definitive, since a decrease in the concentration of the pheromone in the environment can transform a male gametophyte into a hermaphrodite gametophyte.

Maternal Sex-Determination Systems

In a number of animals, the sex of the offspring depends on the mother. This category, *maternal sex determination*, in some ways cuts across the previous systems, because here genetic and environmental factors can interact in different ways and to different degrees in establishing the sex of the individual.

Maternal sex determination can take very different forms. At one end of the spectrum there are some forms of strict genetic determination, where however it is not the individual's genotype that determines its sex, as in ordinary genetic sex-determination systems, but that of its mother. At the other extreme there are cases in which the phenotype or a particular physiological condition of the mother determines the sex of the offspring. In this case, the mother is the source of the environmental signal that determines the offspring's sex.

An example of maternal sex determination of the genetic type is found in two fly species, *Chrysomya albiceps* and *C. rufifacies*. In these insects there are two types of females: androgenic females, which generate only male offspring, and gynogenic females, which produce exclusively female offspring. Gynogenic females are heterozygous (*Ff*) at a gene locus *F* which encodes a maternal factor that accumulates in the eggs and determines the female sex condition in the zygotes that originate from them. At the same locus, males and androgenic females are homozygous recessive (*ff*) and do not produce the maternal factor. The sex of an individual is therefore established at fertilization. However, this is not the fertilization event that produced the zygote from which the individual develops, but the fertilization event that produced the zygote from which its mother developed! In contrast, an example of maternal sex determination of the environmental type is provided by the small polychaete *Dinophilus gyrociliatus*. This produces both large eggs, which will develop into females, and small eggs, from which males will grow. The development of sex characters is regulated by genes whose expression is modulated by the quantity of the trophic resources present in the egg.

However, the two categories have some overlap, showing the interaction between genetic and environmental factors in sex determination. In *Heteropeza pygmaea*, the cecidomyiid midge we already met in Chapter 2, the sex (X0 male, XX female) of the offspring generated by parthenogenesis depends on the nutritional condition of the mother. If this is good, females will be born, males otherwise. In response to the nutritional conditions, a factor produced in the mother's brain is secreted into her circulating fluids, and from there it reaches the ovaries. In the gonad, this factor determines the course of oogenesis, with or without the loss of an X chromosome. Note that, although the primary cause of the sex of an individual is environmental/maternal, the implementation of this signal is genetic (X0/XX). Another case of this kind is provided by certain parasitoids. These are insects that lay eggs within or on the body of another arthropod and whose larvae feed and develop in or on it, eventually killing the host, while the adult is free-living. In many species of parasitoid hymenopterans, the mother is able to control the fertilization of the eggs by the sperm she has in store since her only mating, thereby determining the sex of the offspring, even if the most proximate cause of sex determination in the latter is the haploid or diploid condition of its genome. Fertilized eggs (which will develop into females) are preferentially laid into hosts of large size, whereas the unfertilized eggs (which will develop into males) are generally laid into smaller-size hosts.

Maternal sex determination is also observed in species with heterogonic life cycles, with parthenogenetic and amphigonic reproduction alternating seasonally through the year. In the water flea *Daphnia*, in response to specific environmental signals such as the shortening of the day length or an increase in population density, parthenogenetic females switch from the exclusive production of females to the generation of males and females, which in these crustaceans are all genetically identical to the mother. Environmental signals are received by the parthenogenetic mothers and converted into endocrine signals (juvenile hormone). These have an effect on the maturation of the eggs, which results in eggs that will develop without being fertilized.

In most aphids with heterogonic cycles, in autumn, in response to environmental signals anticipating the approaching adverse season, a generation of parthenogenetic females is followed by a generation of females, which, also by parthenogenesis, produce individuals that will reproduce by amphigony. Aphids have an X0 sex-determination system, but in parthenogenesis the sexual genotype is established under the control of the maternal hormones through the peculiar behaviour of the chromosomes during the maturation of the eggs. Some eggs will lose an X chromosome and will develop into males, while others will keep it and will develop into females. Since in males (X0) only sperm possessing an X chromosome are viable, while sperm without it degenerate, only females (XX) are born from fertilization, the founders of the generations of parthenogenetic females of the following season.

Notes on Sexual Differentiation

Sex determination results in the developmental processes of sexual differentiation, which implement the development of all the characteristics of a given sex condition. Like sex determination, sexual differentiation is a complex process, often including many regulative steps.

In mammals, for instance, we distinguish between *primary* and *secondary sex determination*, which, despite the names, are actually two steps in the process of sexual differentiation. Primary sex determination consists in the differentiation of the gonad into an ovary or a testis. It is a developmental process that proceeds from an embryonic, not sexually committed precursor of the gonad. As we have seen, the Y chromosome is a key factor for the development of the testis, and in its absence the gonad develops into an ovary, although a second X chromosome is also necessary for the complete formation of the female gonad. Secondary sex determination concerns all sexually dimorphic phenotypic characters except for the gonads. These characters include parts of the reproductive system other than the gonads, such as gonoducts, accessory glands and external genitalia, and secondary sexual characters, that is, features outside the reproductive system such as colour and colour patterns or anatomical structures specific to one or the other sex (Chapter 4). These characters are usually determined by hormones (typically, steroids) and other factors secreted by the gonads. In the absence of gonads or their products, a female phenotype develops. Ovaries produce oestrogen, a hormone necessary for the development of the female reproductive apparatus, whereas testes produce the anti-Müllerian factor, which inhibits the formation of female organs, and testosterone, a hormone that masculinizes the fetus, inducing the formation of several components of the male reproductive apparatus. However, even in mammals, some secondary characters are directly controlled by factors produced by genes on the sex chromosomes, rather than by the circulating hormones. In the wallaby *Notamacropus eugenii* the marsupium and the mammary glands in the female and the scrotum in the male begin to form before the gonads, at a time when those sex hormones are not yet in circulation. In the house mouse (*Mus musculus*) there are significant between-sex differences in gene expression before gonadal differentiation takes place.

In animals other than vertebrates, sexual differentiation depends less strictly on circulating hormones, or not at all. In *Drosophila*, and in insects in general, each cell is sexually determined by its own genotype, in a way substantially independent of signals from neighbouring cells or gonadal secretions. Although there are exceptions, this general principle of sexual development is common to many animals.

Sexual differentiation not regulated at a systemic level explains the occurrence of abnormal individuals with a mix of phenotypically male and phenotypically female body parts, called *gynandromorphs*. In some of them, male-type tissues and female-type tissues are distributed in a patchwork throughout the body, but they often occur in a pattern with some kind of symmetry as, for instance, in bilateral gynandromorphs, where one side of the body has male characters, the other side female. Gynandromorphs are found in nature, but can also be induced experimentally. Many cases have been described, mostly among insects, spiders, mites, decapod crustaceans and birds. The general interpretation is that gynandromorphs are genetic mosaics, a mix of cells with male and female karyotypes, as in the case of the zebra finch mentioned in Chapter 1, which had a mix of ZZ and ZW cells. During development, if something goes wrong in the specification of a cell's sex identity, this will affect all the cells that derive from it by mitosis. The geometry of the anomaly will depend on the time and location of the mutation, in the specific developmental context. But there are other possibilities, as a gynandromorph can also stem from abnormalities occurring later in development, either at the level of the processes of sex determination in different tissues or during sexual differentiation. This explains some rare forms of gynandromorphism recorded among vertebrates with gonad-dependent sex differentiation, for instance, among snakes.

At a descriptive level, gynandromorphism can be distinguished from *intersexuality*, where abnormal individuals have sexual phenotypic traits intermediate between male and female. Intersex individuals are quite common in some groups of insects and crustaceans. In the moth *Lymantria dispar*, crosses between different geographic strains result in the generation of intersexes that exhibit an intermediate uniform phenotype, with wings uniform in colour but with different tones from female-like light brown, to medium, to male-like dark brown. Since all these individuals have female karyotype (ZW), the cause of the intersex phenotypes is credited to the incomplete masculinization of 'genetic females' due to a particular allelic combination produced by the cross between strains, that affects the normal expression of the sex-determining genes.

However, a neat distinction between gynandromorphism and intersexuality, although often useful for descriptive purposes, is not always possible or sensible. For instance, zooming in towards finer detail in the description of certain intersex patterns, a gynandromorph pattern can pop up, and vice versa. Or a developmental anomaly can result in both mosaic male/female features and intermediate phenotype features. For example, gynandromorph-like and intersex-like patterns coexisting in the same individual have been recorded in the spiders *Entelecara flavipes* and *Micrargus herbigradus*. These are good reasons for not stressing the distinction between the two phenomena too strongly.

Typological Versus Populational Thinking in Sexual Characters

Consider the scattergram in Figure 7.3. What does it mean? A sensible answer would be that it depends on what the points represents and what *X* and *Y* stand for. Trivial as it may seem, this answer actually conveys an important message.

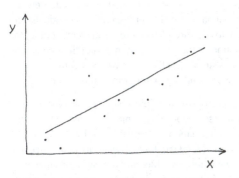

Figure 7.3.   A scattergram of the relationships between two variables *X* and *Y*. The straight line through the points is the linear function that best fits the distribution of data.

Suppose that the points in the plot are repeated measures of, say, pressure (Y) and temperature (X) of a closed gas system at different temperatures and constant volume. If so, the line drawn through the points, mathematically calculated from the configuration of points, would possibly represent the physical law that links the two variables together (in our example, Gay-Lussac's law of pressure–temperature for an ideal gas). The scattering of the points is largely due to measurement error and uncontrolled extraneous factors. The line allows us to see the 'real' law hidden in the cloud of scattered points. In this case, the physical law ($Y = kX$) says that the two variables are directly proportional: when X doubles, so does Y, and for a unit increment in X there is a corresponding k increment of Y.

However, if the points are instead independent observations on as many individuals of a population, and X and Y are two features measured on them, say length of the right wing and length of the right foreleg in a sample of flies, our interpretation of the scattergram will be completely different. Probably, the scattering of the points around the line is only minimally due to measurement error, and rather strongly depends on the fact that individual flies have different shape, at least in terms of their combination of X and Y. What we can say is that there is a statistical association between X and Y, represented by the line ($Y = kX$) we have calculated in exactly the same way as for the physical phenomenon, which indicates a tendency of the two variables to vary in a correlated way. In the specific case, individuals with longer wings tend also to have longer forelegs, and vice versa. But we also note that there are individuals with long wings that have legs shorter than other individuals with shorter wings. Here, contrary to the case of pressure versus temperature, the cloud is the 'real' representation of the phenomenon, while the line is only a useful abstraction. This line is simply *not* the place where ideally the points should lie in the absence of any perturbing factor or measurement error.

In biology, when we take measures of multiple characters in a sample of a population and study their association, this is not to extract what is the 'typical', 'ideal', or 'normal' combination of the values of these characters. This would be to apply *typological thinking*. Rather we try to summarize and quantify the association between these variables and the variations that exist. Acknowledging variation as a fact of nature is referred to as *populational thinking*.

The contrast between typological and populational thinking applies perfectly to the question of the sexual characterization of an individual. There is indeed a statistical association between a number of sex-related characters. For instance, males in our species tend to be heavier than females, with a lower-frequency voice and with the skin more extensively covered in hair. However, this does not mean that there is a typical male character combination, and that any single individual can deviate more or less strongly from it. The observed individual variation in the expression of sexual characters is not the result of the application of a rule with the addition of some degree of error. On the contrary, the relationship between different sex characters derives from the actual variation in sex characters in the population.

But let's focus on another effect of typological thinking on this subject. Since in most living species there are two sexes (two kinds of gametes), one could be inclined to think that characters related to reproduction are highly polarized, either male-type or female-type, so that there is a single species-specific way to be male or female. However, that is simply not the case. For many sex-related characters, different from species to species, there is a lot of heritable variation within each sex. There is variation in size and shape in both males and females, in male aggressiveness in defending a harem, in female propensity to provide care to offspring, in pollen or seed production, and so on. We will give two examples for characters involved in sexual selection (Box 4.1). One of the two main modes of sexual selection is female choice. A male's opportunities to mate and thus his prospective reproductive success depend on the female's fondness for some of the male's specific features. However, female mating preferences are not a species-specific, stereotypic trait. For instance, in guppies there is substantial variation among females both in the order with which a female ranks the prospective mates and in female 'choosiness', the effort she is ready to invest in mate assessment.

As for males, there can be more than one successful strategy in male–male competition, the other of the two main modes of sexual selection. Males of the isopod crustacean *Paracerceis sculpta* have three genetically determined morphs. The largest morph dominates the other morphs by guarding several females in a harem. Males of the intermediate-size morph are similar in size and shape to females, so that by mimicking the latter they can go unnoticed by the dominant males, enter the harem, and mate. The very small males of the third morph use furtiveness and agility to sneak into the harem and gain access to the females. Despite their striking differences in size, shape and behaviour, the three morphs have equal reproductive success.

Summing up, there is not only an extraordinary diversity in reproduction modes across the tree of life, as we have emphasised throughout this book, but also a sizeable variation in characters related to reproduction (and their association) among the individuals of a species or population. Variation is said to be the 'fuel' of evolution, as it provides the 'rough material' for evolutionary sorting processes like natural selection and random drift. Sexual characters are no exception.

In this chapter we have seen how sexual characters develop (Table 7.1). In the next, we will see how they are used.

Sex-determination system
- genetic
  - chromosomal (XY, X0, ZW, Z0, UV, multiple)
  - haplodiploid (female diploid, male haploid)
  - others (males and females with distinct alleles at specific scattered loci)
- environmental
  - temperature-dependent
  - through interaction with conspecifics
  - others (day length, water acidity, nutrition, parasites)
- maternal
  - genetic
  - environmental

Sex-determination mechanism
- with dominant sex chromosomes (e.g. XY in mammals)
- with genetic balance (e.g. XY in *Drosophila*)
- others (with environmental sex determination)

Table 7.1 Summary of the most common sex-determination systems and mechanisms.

8 Widening the View: Reproductive Strategies

Resources for Reproduction and How to Spend Them

In the course of their lives, organisms spend time and energy on a number of activities and functions, of which reproduction is only one – think of growth, defence against predators and pests, and others. How many resources are used for reproduction, how much time is devoted to it and how this time is distributed over the course of life are all elements that characterize the different *reproductive strategies*. From an even wider perspective, in those organisms that at certain times in their lives can opt for one or another reproductive mode (e.g. sexual or asexual reproduction, as in many plants and many marine invertebrates), a reproductive strategy includes also this reproductive policy.

There are many items in the expenditure column of the budget, not all easy to quantify. Focusing here exclusively on biparental sexual reproduction, we survey some of these costs. These different forms of *parental investment* contributing to the survival of the offspring are implemented before, during or after the moment at which the products of reproduction (gametes, embryos, larvae or juveniles) are released.

In animals, the first cost to be considered is the production of gametes, especially eggs. Although less numerous (often, by several orders of magnitude) than the sperm cells produced by a male or hermaphrodite individual of the same species, eggs usually represent a much more conspicuous cost, because of the yolk usually stored in them. But there are exceptions, like the eggs with very little or no yolk of many marine invertebrates (with a diameter in the order of one-tenth of a millimetre), or those of some parasitoid insects, which are laid in the egg or larva of another insect, and are thus literally surrounded by food on which the larva can feed from hatching. An egg of the tachinid fly *Clemelis pullata*, one of these parasitoids, measures just 0.03 × 0.02 millimetres.

We should not, however, underestimate the cost for some animals of producing sperm and the materials that often accompany them. Sperm production can be a significant expense in the case of promiscuous mating systems (Chapter 5), so that the strategy adopted for distributing the gametes between different and frequent episodes of insemination (*sperm allocation*) can be an important aspect of the breeding strategy, as in some fishes. Furthermore, the production of male gametes is usually accompanied by the secretion of a *seminal fluid*, which together with the sperm makes up the *semen* (or *ejaculate*). This fluid helps or allows sperm cells to reach the eggs, but can also have other functions, such as nourishment for the female, or limiting the risk of competition from another male's sperm. Some insects, such as the bed bug *Cimex lectularius*, produce large amounts of sperm and/or seminal fluid that to a large extent ends up nourishing the female after being absorbed into specialized structures. The *spermatophore* of the Mormon cricket (*Anabrus simplex*), a large insect up to 8 centimetres in length, is enormous, weighing a quarter of the male's entire weight, and is mostly food for the female. In many species where a female can mate with multiple partners a male can inactivate the sperm of previous mates through chemicals contained in his seminal fluid. This is known for example for the fruit fly *Drosophila melanogaster*. In scorpions, sperm is transmitted in spermatophores that break apart, allowing the sperm to enter the female's genital tract, while a part of the envelope forms a persistent plug on her genital opening. In many rodents, but also some primates and reptiles, the ejaculate contains substances that stiffen after mating, creating a *sperm plug* that for a while can make insemination from other males difficult. The production of gametes and seminal fluid can therefore be loaded with additional functions, and thus costs.

Another cost for the mother is provisioning the offspring with food, a behaviour that often extends in time well beyond the production of mature eggs, as seen in the various forms of matrotrophy described later in this chapter. Reproductive costs typically faced by the male are the 'bridal gifts', very common among birds, but also found among insects, such as the prey enclosed in a silk wrapper that the male of many empidid flies offers to his partner. The ultimate gift is the male himself being eaten by the female during mating, as in some mantises and spiders.

Other costs can be borne by either or both parents, depending on the species. Very widespread but not universal expensive activities are the exploratory behaviour in search of a partner, nest building, excavating burrows, producing cases or other devices for the protection of the eggs. Another cost may arise from the lack of food intake that must be endured by the parent that takes charge of incubating the eggs. This is generally the female, but many birds, especially seabirds such as petrels and albatrosses, share the duties equally, and in the emperor penguin (*Aptenodytes forsteri*) it is the male who fasts for the entire duration of incubation (62–67 days) in the extreme environmental conditions of the southern winter on the margins of the Antarctic continent.

For the sporophyte of the seed plants, the dominant generation in their haplodiplontic cycle, we must take into account both the costs related to pollination and those connected with the production of seeds and fruits, including their dissemination. The less noticeable costs sustained by the gametophyte are presented in the last section of this chapter. For plants that rely on wind pollination (Chapter 5), whose efficiency is low, the greatest cost is in the production of large amounts of pollen, of which only a small fraction stands any chance of reaching another flower of the same species. For plants with animal-assisted pollination, the production of large and colourful corollas, or showy specialized leaves, such as the involucral bract surrounding the inflorescence in the arum family, may represent significant costs. The same goes for nectar and perfumes, as well as the additional pollen that is consumed by pollinators. At the end of this chapter, we will examine the costs associated with seed dispersal, including the abundant fleshy pulp of berries and structures such as the wings of the dry fruit of maples or the seed heads of dandelions and thistles.

Fecundity

An element of primary importance in any reproductive strategy is *fecundity*, which is a measure of the number of descendants.

Actually, the term fecundity has both a qualitative and a quantitative meaning. The first refers simply to the ability of an individual to reproduce. For many organisms, including our species, this capacity is also called *fertility*. In the second sense, fecundity is instead a measure of the numerical abundance of descendants, and it can be applied to individuals but also to species. Fecundity can refer to a single reproductive episode, to a single reproductive season, or to the whole life of an individual (*lifetime fecundity*). The distinction between fecundity in the single reproductive season and fecundity over the whole life of the organism is clear for many animals and plants, where separate reproductive seasons occur in an individual's life. In other cases, however, such a distinction becomes meaningless, either because the organism lives for only one reproductive event, or because the reproductive phase can be prolonged for a very long time with uninterrupted production of offspring. Exemplary, in this regard, is the serranid fish *Diplectrum formosum*, a simultaneous hermaphrodite that releases gametes every two days during an adult life span extending up to eight years.

How to measure fecundity is not the same for all. It depends on the reproductive modes of the organism, its life cycle and the aspects of its biology (physiology, ecology, evolution) we want to investigate. In asexual reproduction, fecundity is generally measured as the number of propagules produced with a reproductive event, in a reproductive phase, or within a certain time interval. In most animals, the usual numerical disproportion in favour of male gametes suggests that we should measure fecundity based on the number of eggs produced, which are the numerically limiting resource. The female of the splendid fairywren (*Malurus splendens*), an Australian passerine bird, lays six eggs at a time, while the male always has eight billion sperm cells at his disposal, potentially suitable to fertilize as many eggs. In the coho salmon (*Oncorhynchus kisutch*), the corresponding figures are 3,500 eggs to 100 billion sperm cells. Fecundity estimates in animals sometimes refer to the number of eggs produced irrespective of whether they are fertilized, an acceptable choice in cases where it is actually possible that most of the eggs produced by an individual will be fertilized. However, in many animals, as in mammals, the maximum number of offspring generated by a female is much lower than the total number of eggs produced: in this case, it seems reasonable to measure fecundity based on the number of offspring born. In seed plants, the most common measure of fecundity is the number of seeds produced, while for mosses and ferns, fecundity is generally quantified as the number of spores produced.

Ecologists distinguish as *r* and *K* two extreme reproductive strategies corresponding to two opposite ways in which the expected reproductive success can be maximized: in the *r-strategy*, many small eggs (or many small seeds) are produced, and no parental care is provided, whereas in the *K-strategy* fewer offspring are generated, but supported by sizeable nutritional supply, prolonged parental care, or both. The reproductive strategies actually observed in nature are found on a continuum between these two extremes. This rules out both the extreme economy of producing very few eggs with little yolk (or very few seeds with little endosperm) and abandoning them to their fate and the extreme burden of producing many eggs or many large seeds and/or providing long, demanding parental care to very numerous offspring. The first would be a losing strategy due to the very low probability of survival of the offspring, the second is simply not practicable.

The lowest number of offspring per female may be expected among viviparous animals, especially those, like most mammal species, in which parental care extends significantly beyond the gestation period. However, the record for the lowest fecundity is possibly found among the insects.

In the vast majority of holometabolous insects (the group where the larva metamorphoses into an adult with very different shape and habits, through a mostly inactive pupal stage), feeding is mainly, or even exclusively, the larva's responsibility. Adults may take additional food, as butterflies do on flowers and female mosquitos by piercing through our skin, but their main business is mating and egg laying (and sometimes providing further resources to the offspring; see *Parental Care in Animals*, below). However, a few insects deviate dramatically from this general pattern. One example is seen in some species of the beetle tribe Leptodirini, blind wingless insects adapted to life in deep caves, one of the least hospitable and most resource-poor environments on Earth. These beetles lay a very small number of eggs, and in the most extreme case just one at a time, of enormous size. The larva that hatches from it does not take any food and does not moult to a more advanced active larva (most beetles develop through three larval stages), but a few days after hatching it builds a solid chamber of sand and gravel within which it will become a pupa and eventually an adult. In tsetse flies, the female matures only one egg at a time, which develops inside the body of the mother, feeding on secretions produced by her. When it is eventually released, the voluminous larva is ready to metamorphose, without taking on further food. Similar is the larval development in other families of Diptera whose adults are all parasites of mammals or birds.

Most sessile marine invertebrates, such as sponges and corals, produce a very large number of eggs and release their gametes (female as well as male) into the water, where fertilization takes place (these are examples of an *r*-strategy). The fecundity of many fish species is also very high. For example, a large grouper can produce in its lifetime tens of millions of eggs. Large numbers of eggs per clutch are produced by crabs. In some *Cancer* species, the total number of eggs produced by a female during her lifetime is estimated to be more than 20 million. By contrast, some small marine invertebrates, including many interstitial species, inhabiting the spaces between individual sand grains of the sea floor, produce only a few large eggs (a *K*-strategy).

Among the insects, the queen of the African driver ant *Dorylus wilverthi* lays up to 3–4 million eggs every 25 days. In most birds the individual clutch rarely exceeds 4–5 eggs. A major exception are gallinaceous species (Galliformes), among which the quail and the wild turkey lay on average more than 10 eggs, the Daurian partridge up to 19. Slightly more than 10 eggs on average are also produced by several ducks and geese (Anseriformes) and passerines, for example tits. On the other hand, many small birds, such as the hummingbirds, lay only two eggs at a time, while others, for example many seabirds, many doves, some swifts and several nocturnal birds of prey, lay just one egg. In mammals, litter size varies from one (all cetaceans, the chimpanzee, the mountain zebra) to one or two (bats and giraffes) to 4–6 in the European hedgehog and the house mouse, to up to 12 in the wild boar. A maximum of 28 has been reported for the naked-mole rat, and the tailless tenrec (*Tenrec ecaudatus*), from Madagascar, can produce as many as 32 in a single litter (served by up to almost 30 nipples!).

As for plants, in the case of spermatophytes the most obvious index of fecundity is the number of seeds produced, while for terrestrial plants with gametophytes not dependent on sporophytes (bryophytes and pteridophytes), fecundity is expressed as the number of spores produced. In flowering plants, the number of seeds yielded in a season depends both on the number of developed fruits and on the average number of seeds per fruit. The latter number spans six orders of magnitude, from one seed, as in the coconut and the peach, to almost four million in *Cycnoches* orchids. Most orchid species have very small seeds and no endosperm: seed germination and the nourishment of the embryo up to seedling growth depend on symbiotic interaction with a fungus. The fungal partner supplies nutrients required by the plant before it can get nourishment through photosynthesis. Most orchids associate with only one fungus species and vice versa, but there are exceptions. There are about 5–10 seeds in an apple, 120 in a papaya and 200 in a watermelon. The fruit of the invasive tree of heaven (*Ailanthus altissima*), native to China, has one seed per fruit, but can produce more than a million fruits per year.

A wide range of reproductive strategies along the r–K continuum is also found in the mosses: a mature sporophyte of *Archidium alternifolium* releases only 16 spores, but there are 50 million in a mature sporophyte of *Dawsonia lativaginata*.

Lifelong Temporal Distribution of Reproductive Effort

One defining aspect of a species' reproductive strategy is how reproduction is distributed in time over an individual's life.

In many species, reproductive activity is restricted to one short period of the organism's life and is followed very soon by death, especially in male animals. The scientific name for the delicate, flimsy insects known as mayflies is Ephemeroptera, based on the scientific name of the first-named genus in this insect order. *Ephemera* means, literally, 'for one day,' but the adult phase of a mayfly's life is mostly even shorter: just a few hours to mate and, if female, to lay eggs. In these insects, death is already waiting after this fleeting moment, only five minutes in the females of *Dolania americana*. The same is true of some very much larger animals, such as male squid, which discharge all their spermatophores at once, and then die. Uniquely among the terrestrial vertebrates, the chameleon *Furcifer labordi* has an annual life cycle, so they reproduce once, after about seven months as embryos and only 4–5 months as active animals, which makes their life cycle resemble that of many insects. The presence of just one reproductive event in an organism's life (*semelparity*) is not limited to short-lived animals or plants. This is indeed a characteristic trait of some species of agave, the eels, some species of salmon, as well as ticks and other robust arthropods such as the South American scorpion *Bothriurus bonariensis*.

In other cases, the segment of life during which an organism reproduces extends without appreciable interruptions over months or years. In long-lived organisms that reproduce seasonally throughout most of their life, the reproductive output generally decreases year after year, an obvious sign of progressive senescence (Chapter 1). However, there are exceptions. In *Dioscorea pyrenaica*, a herbaceous plant belonging to the same family (and genus) as the yam, 260-year-old individuals have the same reproductive capacity as plants of the same species at the start of the adult phase.

In still other cases, the individual has multiple reproductive seasons, more or less circumscribed in time and separated by long periods of reproductive rest. In some ctenophores and polychaetes, a first reproductive season in the larval stage is followed by a second more conventional one as an adult, while the males of some millipedes have two successive reproductive seasons separated by two moults, between which the animal's copulatory structures regress and reproduction is suspended.

On a large scale, these different reproductive strategies have a great influence in determining the resources allocated to a single reproductive episode.

The distribution of reproductive activity over time may also reflect more specific adaptive strategies. Many tree species that are dominant in some forest communities and produce seeds and/or fruits particularly favoured by local primary consumers alternate, more or less regularly, between several years of ordinary seed production and single years of particularly abundant production. For example, at temperate latitudes this occurs in the European beech (*Fagus sylvatica*) every 4–5 years, with even more exceptional fruiting every 10–15 years. During the years of abundant fruiting, the production of seeds and fruits tends to saturate the herbivores' demand, thus increasing the chances of survival of the seedlings of the new generation. At the same time, the irregular occurrence of these exceptional years does not allow herbivores to evolve specific adaptations to exploit them fully.

The same principle of demand saturation may also explain the extraordinary synchronism in the flowering of most woody bamboo species. All the plants of a particular species bloom at the same time, regardless of where they grow and independent of the climatic conditions to which they are exposed, and die at the end of the flowering period. This is even more extraordinary in consideration of the fact that this extends to all the plants generated asexually by rhizomes or cuttings from a given generation of bamboos born from seed. Asexually generated plants will bloom (and die) together with those that have grown from seed, as if they were born with the age of their sexually generated progenitor (a form of clonal senescence; Chapter 1). This mass flowering is repeated at rather regular intervals ranging from a few years in several species up to 60 in *Phyllostachys nigra* and over 120 in *Ph. bambusoides*. In the latter species, the current generation is expected to flower around year 2090.

In addition to the number of breeding seasons in an individual's life, another aspect of the temporal distribution of the reproductive effort is the age at which an individual begins to reproduce sexually. In animals, this age varies greatly from species to species, even disregarding the cases in which the individual reproduces when still a larva or a juvenile (paedogenesis; Chapter 6). Within mammals, guinea pigs and other rodents can reach sexual maturity in less than two months, while many ungulates take about a year, and primates (including humans) and dolphins require more than 10 years. Some whales take even longer, about 23 years in the bowhead whale (*Balaena mysticetus*). In mammals and birds there is a direct relationship between the age at the onset of reproductive maturity and the average size of the adult animal. In species with marked sexual dimorphism, especially in body size, males and females often reach reproductive maturity at different ages. In a number of fishes, females mature later than males (often as male-first sequential hermaphrodites; Chapter 4), while the opposite is observed in birds and mammals.

In seed plants, the onset of reproductive maturity signals the entry into the *adult reproductive phase*, characterized by the acquisition of competence for sexual reproduction through the development of reproductive organs such as strobili or flowers. In fast-growing herbaceous species, the stages of development that precede the onset of flowering are often very short, as little as one month in some populations of the shepherd's purse (*Capsella bursa-pastoris*), a common plant of the cabbage family. However, the juvenile, pre-reproductive phase is sometimes very long even in herbaceous plants, for example 8–16 years in eastern Swedish populations of the sanicle (*Sanicula europaea*), a member of the carrot family, and 13–14 years in the dandelion *Taraxacum stevenii* from the Caucasus. In a number of tree species the onset of first flowering may require more than 20 years, as in many oaks. In long-lived perennials, the age at the onset of reproductive maturity may vary within one species even more than in animals, depending on the environmental conditions to which the plant is exposed during the pre-reproductive phases of development. Perennial plants often continue to produce new reproductive organs from tissues capable of growth and differentiation in shoot tips (apical meristems), and therefore adverse environmental conditions, pest attacks, damage by grazing animals or other external causes can strongly affect the continuity of reproduction during the adult phase.

Finally, times of reproduction are not only very different in different species, but also sometimes difficult to calculate, for example when phases that usually occur in rapid succession and in a precise sequence are instead interrupted, for even a long time, or even occur in reverse order. For instance, at the time of the penetration of the sperm, the female egg is not always a 'true egg', that is a germ cell that has completed meiosis. In amphibians and mammals it is fertilization (of what is not yet an egg) that triggers the completion of the second meiotic division. A prolonged arrest of embryonic development is a rule during the gestation of some mammals. For example, in roe deer (*Capreolus capreolus*) mating occurs between July and August and fertilization follows immediately, but the implantation of the embryo on the uterine wall does not occur before the end of the year. This *embryonic diapause* is also known in a number of other mammals, including kangaroos, rodents, seals and bears, but also in some reptiles and fishes. It makes sure that birth takes place at the most favourable time of year for the offspring, regardless of when fertilization took place. In many conifers and other gymnosperms, such as cycads and ginkgo, there is usually a long time between pollination and fertilization (Chapter 5), and the ovule begins to develop into seed long before fertilization is complete. In the Scots pine (*Pinus sylvestris*), for example, female cones first become visible in the spring, but they will not be receptive to pollen until late May or early June. Shortly after pollination, the pollen grain germinates, sending out a short pollen tube. The germinated pollen remains dormant for a little more than a year after pollination, when it resumes growth and fertilizes the ovule.

Mothers, Eggs and Embryos in Animals

A traditional classification based on the relationships between the egg (or embryo) and the maternal body divides animal species into the three classes of oviparous, ovoviviparous and viviparous. However, a more recent classification considers separately the state in which the offspring leave the maternal body, either while still being eggs (oviparity) or in a more or less advanced stage of post-embryonic development (viviparity), and the way in which the mother provides them with nourishment, either by storing yolk in the egg (lecithotrophy) or by transferring nourishment to the offspring at a later stage (matrotrophy) (Figure 8.1).

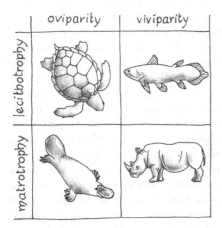

Figure 8.1.   Two-way classification of the different strategies in the relationship between a mother and her offspring's early development in animals, with an example for each combination: the loggerhead turtle (*Caretta caretta*), the coelacanth (*Latimeria chalumnae*), the platypus (*Ornithorhynchus anatinus*) and the black rhinoceros (*Diceros bicornis*).

This double classification accommodates the conditions of ovoviviparity in the traditional sense (viviparity associated with lecithotrophy), but also some apparently atypical cases of oviparity associated with matrotrophy, as exemplified by the monotremes. These animals – the platypus and the echidnas – are the only oviparous mammals, but they are also matrotrophic, albeit in a very singular way. In other mammals, during intrauterine life nutrients pass from the mother to the fetus through the placenta, while in monotremes the embryo receives a substantial nutritional contribution through the thin shell of the egg, before this is laid.

In *oviparous* animals, eggs may be released all at once or in a number of separate events, in groups or individually, sometimes over a long time span. Eggs released in the water may be covered by gelatinous coatings of various origin and thickness (e.g. most water snails), or more rarely enclosed in protective cases like the parchment-like cocoons of a number of leeches and some cartilaginous fishes such as skates and dogfishes. In terrestrial environments, both land vertebrates and several lineages of arthropods have evolved measures to avoid rapid desiccation of the eggs. In addition to different forms of incubation (see below in this chapter), the problem has found a number of solutions. Many terrestrial animals return to water to release their eggs. This is the strategy of most amphibians and of several groups of insects, among them mayflies, stoneflies, mosquitoes and some dragonflies. For many other insects, instead, water loss is prevented by laying the eggs within the tissues of another organism. This may be a victim (generally the egg or the larva of another insect) that will also provide nourishment to the larva (as in many parasitoid wasps), but in other cases protection is given by the tissues of a plant even if the insect is carnivorous. Eggs are forced, one by one, into green stems (less frequently, leaves) thanks to the ovipositors of damselflies, aeshnids (a family of large dragonflies), sawflies and many crickets and grasshoppers. A third solution is the production of shells and/or protective cases, the latter sometimes used to cover the single egg (e.g. stick insects), otherwise to protect a whole batch of eggs (e.g. cockroaches and mantises). A last series of adaptations involves the active behaviour of the mother (more rarely, the father or both parents) that modifies the environment by digging a den or a gallery or adapting an existing one, or building a nest to house the eggs – where the offspring may additionally benefit from parental care extending beyond oviposition.

Wasps, bees and ants use a huge variety of shelters or true nests for the protection of their offspring, to which, depending on the species, they also offer more or less prolonged and differently sophisticated care. The more complex behaviours are found, in general, in the so-called *eusocial* species, where individuals are organized in castes, not all of them being able to reproduce. In the wasps, eusocial forms such as hornets (*Vespa* spp.) build nests using a sort of cardboard obtained by chewing wood fibres. Among the solitary forms, many make mud nests instead. As in the ants and the honey bee, the larvae of eusocial wasps are fed by sterile adult individuals (workers), while the solitary wasps leave in their nests, together with the egg, the food that the larva will feed on, represented in this case by prey, paralyzed by the mother with poison injected by a stroke of the sting. A number of solitary bees and wasps build nests of unusual workmanship. Several solitary bees exploit habitable spaces such as the hollow stems of plants with soft pith, and divide the cavity into individual cells, one for each egg. A wide variety of materials is used for this purpose: bees of the genus *Megachile*, for example, cut out circular or oblong pieces from the leaves of roses and other plants, while *Osmia papaveris* uses scraps from the red petals of poppies, and her relatives *O. bicolor* and *O. rufohirta* nest inside empty snail shells, dividing the internal space into several cells with chewed plant material. Other wild bees, such as *Anthidium strigatum*, build nests with resin collected from pine trees. Inside their nests, all bees accumulate food for the larvae that will hatch. This food typically consists of pollen; only honey bees (the European *Apis mellifera* and its close relatives) are able to produce honey.

In *viviparous* animals the embryo develops within the reproductive system, the body cavity or the tissues of the parent until the offspring is released as a larva or a juvenile. This is different, in principle at least, from *incubation* (see below in this chapter), where the offspring, released as a zygote, embryo or even in a post-embryonic stage, is maintained in close contact with the surface of the mother's body or inside a fold or a pouch, sometimes but not always specialized for this function. These behaviours, which in some cases extend beyond hatching (and often indicated as *brooding*), verge to a variable degree on viviparity. Viviparity can be associated with matrotrophy. In marsupial and placental mammals some specialized tissues of the mother and the fetus, forming the placenta, allow continuous exchange of fluids between mother and offspring ('placental mammals' is something of a misnomer, since marsupials have a placenta as well; see below). The placenta mediates the transfer of nutrients, but also the exchange of respiratory gases, waste products of metabolism and antibodies. Viviparity accompanied by lecithotrophy (ovoviviparity) is found in scorpions and in many insects and fishes, including about half of the cartilaginous fishes, the coelacanth and a few hundred bony fishes. Isolated cases of viviparity also occur among the amphibians, but the phenomenon is more widespread among the reptiles, with examples among chameleons, iguanas, geckos, lizards and a number of snakes.

There is *lecithotrophy* when the mother's contribution to the nutrition of the offspring ends with storing yolk in the egg. Yolk can be supplied to the egg through the blood or other body fluids. As an alternative, nourishment is provided instead by other cells, through cytoplasmic bridges, or by being phagocytized by the egg. These cells can be abortive germ cells or somatic cells of the ovary. In many animals, some of the eggs end up as a source of nourishment for the other eggs (or for the embryos that emerge from them: oophagy, see below). As we saw in Chapter 4, in many flatworms the yolk is stored in particular cells that are produced in an organ (the vitellarium) distinct from the ovary, but are subsequently associated with a true egg to form a *compound egg*. From a certain perspective, this functional multicellular unit is not much different from the ovule of seed plants, which is actually a functionally integrated egg. Compound eggs are also found in some sponges and other invertebrates.

In *matrotrophic* animals, the mother provides a direct and continuous contribution of nutrients other than the yolk supplied to the egg and the food collected in the surrounding environment for the benefit of the offspring. Matrotrophy can be associated with viviparity or oviparity, and can also extend beyond the end of embryonic development, as in the feeding of offspring with milk by female mammals. In some cases, in spite of the apparent oxymoron, matrotrophy during post-embryonic development can also involve the male parent.

Matrotrophy can be implemented in different ways. The best-known example is *lactation* in mammals. Remember that female monotremes, which are oviparous, also feed their newborns with milk. Duration of lactation is quite variable, from 3 to 5 days in the hooded seal to over one year in the chimpanzee and up to two years in rhinos.

Another form of matrotrophy is the *dermatophagy* of some fishes and amphibians, in which the juveniles feed on specialized skin areas of the parent. The same phenomenon is seen in the so-called crop milk fed to young pigeons and other birds (e.g. flamingos), produced by the degeneration of the cells that line the crop of the adult bird. Father and mother pigeon collaborate in providing this nourishment, rich in fats and proteins, which represents the only food during the first three days of life of the chicks. In the Alpine salamander, the larvae feed on unfertilized eggs (*trophic eggs*) present in the uterus where they are developing (*oophagy*). Similarly, in ants, sterile eggs produced by workers (or by the queen) are sometimes supplied as food to the larvae they care for. In *adelphophagy*, nourishment is provided in the form of sibling embryos present in the female's genital tract, as in the bull shark and in several sea stars. In still other cases nutrients are supplied directly from the liquids produced by some specialized tissues of the mother, and absorbed by the embryo (*histotrophy*), a mode found in a few mites and some sea urchins and sea stars. Histotrophy is probably present in all pseudoscorpions, a group of arachnids resembling miniature scorpions without the characteristic tail terminating in a poisonous sting. Their embryos develop in a brood sac carried around by the mother, who also provides them with nourishment by pouring into the brood sac a nutritive fluid from the walls of the ovary and the oviducts; here, the epithelial cells undergo massive degeneration, but will regenerate after a reproductive event. Transfer of nutrients from the ventral epithelia of the parent and the posterior sucker of the young attached to its body has been demonstrated in some leeches, including two species common and widespread across Europe, *Glossiphonia complanata* and *Helobdella stagnalis*.

Still another form of matrotrophy is *histophagy*, which occurs when the offspring ingest tissues or portions of tissues of the mother, or, in rare cases, even the whole mother (*matriphagy*). This occurs in a few species of nematodes, the larvae of which feed on their mother's internal organs and may complete their development in her body, reaching maturity and even copulating (incestuous mating; Chapter 5) before leaving what remains of their mother.

A particular case of histophagy occurs in some species of caecilians (limbless amphibians), where the newborns feed off their mother's shreds of skin. Matriphagy from the outside of the mother's body also occurs in some spiders, such as *Stegodyphus lineatus*. In this extreme (suicidal) form of provisioning by the parent, the mother first contributes regurgitated food and products of the dissolution of her internal organs to feed the newborn spiderlings, then, once deceased, she is completely devoured by her offspring.

Finally, *placentotrophy* requires close contact or even fusion between some specialized tissues of the mother and those of the fetus, forming a placenta that allows continuous exchange of fluids between them. Characteristic of placental mammals, among which there is considerable structural and functional diversity, a placenta is also found in marsupials, although this has anatomical features different from placental mammals and provides nourishment to the embryo for a relatively short period of time, after which the still fetus-like young move to the pouch to suck milk. In marsupials, where the newborn might be considered a larva (at birth, a kangaroo weighs just 30 millionths of the mother's weight), the intrauterine life is very short, only 13 days in the Virginia opossum. However, the pregnancy of small placentals is also short, for example 16 days in the golden hamster and 21–24 days in the house mouse. At the other end of the scale, gestation is approximately 280 days in humans, and longer still in camels (about 400 days), giraffes (465 days), rhinos (573 days) and elephants (660 days). In spite of the animal's size, the gestation of the blue whale is shorter, estimated at about 11–12 months.

Placental matrotrophy is found also in some representatives of different groups of invertebrates, especially among scorpions, bryozoans and some insect orders.

Parental Care in Animals

Parental care is a specific behaviour of one or both parents that allows the development of the offspring in suitable conditions or at least increases their chances of survival. This care can be provided at different stages of the offspring's life, including both prenatal behaviours such as egg guarding, and postnatal care such as active protection from predators and parasites, and feeding the young.

A primary form of parental care consists in providing offspring with collected food, sometimes reworked mechanically or chemically, with the help of hydrolytic enzymes. In several instances, parents store food in a suitable place, where it remains available to the larvae or juveniles that hatch from the eggs laid in these clusters of food or in the immediate vicinity, as a rule in burrows or tunnels dug by the parent.

Thousands of species of scarab beetles exploit mammal excrement as food, in both the larval and the adult segments of their lives. Besides directly exploiting dung for themselves, either where it is found or in burrows into which they bring sizeable amounts of dung, adult scarabs also care for their offspring, laying the eggs into the food. Three different categories are distinguished. The most primitive are the scarabs that do not displace the dung from where they find it, but simply lay their eggs more or less deep in the mass. Others excavate long vertical tunnels, usually provided with lateral chambers and often branched, in the soil beneath the dung and move the food required by the larvae down into those tunnels. Most spectacular is the behaviour of the so-called rollers, to which the famous sacred scarabs of the ancient Egyptians belong. Male and female, usually acting in tandem, build a spherical mass of dung and roll it some distance from the site where they have found it, and bury it in a subterranean chamber they dig. Eggs are laid in the food mass, and the female of some scarab species remains in the chamber with her offspring until hatching.

In other animals, food will be administered directly to the offspring by the parent or by a helper (the most significant examples of which are the sterile workers of the eusocial insects), often in nests prepared by the parent, alone or with the help of the partner and/or one or more related individuals.

Another form of parental care consists of *incubation* and *brooding*, where the offspring are maintained in close contact with the parent's (most often the mother's) body or inside a fold (such as the mantle cavity of molluscs) or a pouch (as in marsupials, some isopod and amphipod crustaceans, and several anurans), sometimes but not always specialized for this function. In other words, the parent creates a local environment suitable for offspring development. Whatever the details, the relationship with the parent, sometimes extending beyond the end of embryonic development, may consist solely in the protection from environmental injuries, from predators or parasites, or be combined with the direct supply of nourishment (matrotrophy). In birds, incubation may last from 10 days, as in some small passerines, to about 21 days in the chicken and 40–45 days in the golden eagle.

Among the terrestrial vertebrates, anurans are the group with the greatest diversity of incubation patterns, in which one of the parents carries around eggs or tadpoles on itself or within itself. Male midwife toads (*Alytes* spp.) wrap strings of eggs laid by one or more females around their hind legs, and carry them around until hatching. In the Suriname toad (*Pipa pipa*), instead, the female places eggs and tadpoles into deep depressions in the skin of her back and carries them until metamorphosis is completed. In the *Gastrotheca* frogs of Central and South America, the female is provided with a sort of dorsal pouch, which in some species has specialized areas through which exchanges of gas, water and waste products occur between mother and offspring. In Darwin's frog (*Rhinoderma darwinii*), the male, after having guarded the eggs laid by the female among dead leaves for three or four weeks, picks them up when they are close to hatching and hosts the tadpoles in his vocal pocket, where they will remain, feeding on substances derived from the parent's tissues, until they develop into tiny froglets. In *Rheobatrachus silus* the female literally ingests the newly fertilized eggs, and incubation (5–6 weeks) takes place inside the maternal stomach. Actually, we should use the past tense, because this Australian rainforest species became extinct a few years after its discovery, in 1984. This prompts a digression on the unpredictable conse-quences of biodiversity loss on our planet, adding to those we can easily envisage. Investigating how the larvae could escape digestion in the mother's stomach, it was discovered that it was the tadpoles themselves that produced a substance capable of blocking the mother's gastric activity. Researchers predicted potential applications in medicine, for instance, to treat stomach ulcer, a disease that affects millions of people annually worldwide. But it was too late: the alteration of the habitat in which the gastric-brooding frog lived led to its extinction. Any attempts to resurrect it with cloning techniques are, for the moment, unsuccessful.

When parental care is the responsibility of only one parent, this is almost always the mother. However, in addition to midwife toads and Darwin's frog, the care of offspring is entirely entrusted to the father in several animals. Among these are the seahorses, where the male has a ventral pouch, also present in many species of needle fish, which belong to the same family. Paternal care is also observed in some millipedes, in which the male remains coiled around the eggs laid by his partner, until hatching. In the polychaete

Neanthes arenaceodentata the two partners form a common mucous tube inside which they release their gametes. The female dies shortly afterwards, while the male remains with the directly developing offspring until hatching, about 10 days after fertilization, and even for a further three weeks thereafter. The incubation of the eggs is also the full responsibility of the male in some birds. This is often the case in species where polyandry is the norm (Chapter 5), such as jacana and phalaropes. The male emperor penguin spends nine weeks or more incubating through the Antarctic winter (see above), and no less remarkable is the effort of the male Australian malleefowl, who devotes five hours a day for six long months to building and maintaining the gigantic mass of sand and decaying matter in which he accommodates the eggs laid by the female, one at a time. A huge amount of material is reworked each time. The heat required for incubation is released by the decomposition of the organic material of which the nest is largely composed, and the continuous rearrangements performed by the male ensure that the embryos will enjoy appropriate conditions for development until hatching.

In some animals, in addition to the parental care provided by one or both parents, care is supplied by other individuals. These helpers are sometimes individuals specialized for this function, as in eusocial insects (bees, ants, termites), sometimes relatives that do not reproduce, in the current season at least. This is the case in the Florida jay, where young birds often help their parents, feeding their siblings or guarding and protecting them from predators. Another example – one that actually involves non-kin, 'unaware' helpers – is provided by the parasitic behaviour of the European cuckoo and another 60 species of the same family: in these birds the female lays the eggs in the nests of other species, entrusting to them the whole business of incubation and subsequent feeding of the cuckoo chicks.

Provisioning by Plants for Offspring Development and Dispersal

There are interesting parallels between animals and plants in the way that parents care for offspring. We must only take into account that the latter are haplodiplonts, where parental provisions for the development of the gametophyte are granted by a parent sporophyte, while those for the development of the sporophyte are supplied by a parent gametophyte.

In plants with an independent gametophyte (e.g. many algae, mosses and ferns), the investment by the parent sporophyte is usually limited to a modest supply of reserve substances to the spores, and the production of protective coating layers that increase their chances of survival and dispersal. However, there are many cases in which the gametophytes, during more or less extensive phases of development, benefit from the protection and nourishment of the sporophyte that generated them. This is typical of seed plants, although not exclusive to them. Here the female gametophyte (the embryo sac) is retained and nourished by the parental sporophyte; the latter also, although indirectly, ends up for some time providing nutrients to the sporophytes of the next generation that will eventually emerge (see below). The male gametophyte (the pollen grain) develops only thanks to the nutrients received from the parental sporophyte. In angiosperms, a specialized tissue of the anthers, the tapetum, produces a nutritive fluid that contributes to the development and maturation of pollen.

In the bryophytes, plants in which the gametophyte generation is dominant over the sporophyte generation, the nutrition of the sporophyte depends, for a longer or shorter time, on the gametophyte, through cells or tissues specialized for the transfer of nutrients from parent to offspring, similar to the placenta of mammals. In tracheophytes (vascular plants), where the sporophyte is dominant over the gametophyte generation, the female gametophyte is relatively small and has only limited trophic and protective functions, in any case restricted to the embryonic phase of the sporophyte's development.

In ferns and horsetails the embryonic sporophyte is retained by the gametophyte that nourishes it during early development, until it produces the first leaves and the first roots, thus becoming independent. In parallel, the small parent gametophyte generally fades out and eventually dies, a phenomenon analogous to the extreme matrotrophy of some animals.

In spermatophytes (seed plants), the embryonic sporophyte is retained by the female gametophyte, but this, in turn, is supported by the sporophyte of the previous generation. Ultimately, nutrition and protection during the development of a young sporophyte are provided by the tissues of the sporophyte that produced the mother gametophyte. In other words, this is a case of 'grandparental care'. The trophic resources for the early developmental stages of the new sporophyte are provided in various ways. In gymnosperms ovules are gigantic (think of pine nuts; but those of *Cycas* are much bigger, up to 7 centimetres in diameter, the largest in the entire plant kingdom), stuffed with carbohydrates and proteins; in these plants, it is the haploid cells of the female gametophyte that continue to grow and function as a nutritive tissue, the *primary endosperm*, that will eventually be stored in the seed. In angiosperms, instead, the trophic reserves available for the development of the new sporophyte are provided by the *secondary endosperm*, a triploid tissue that derives from the parallel fertilization of another cell of the same ovule by another sperm cell of the same pollen grain (double fertilization; Chapter 5). In some respects, the secondary endosperm could be considered a twin embryo of the new sporophyte, so that this mode of parent-to-offspring transfer of resources through a sibling is somewhat reminiscent of adelphophagy in animals.

In spermatophytes, the parental investment in the young sporophyte, which generally goes well beyond the formation of the egg cell, is modulated mainly through the development of the seed and, in flowering plants, of the fruit. Let's now turn our attention to the development of these two structures.

The *seed* is a composite structure that develops from the ovule in the vast group of plants that make up the spermatophytes. The core of the seed is the embryo of the new sporophyte, issued from the fertilization of the egg (one of the cells of the female gametophyte) by the spermatic nucleus borne by one of the few cells of the male gametophyte. In addition, the seed includes nutritive tissues (primary endosperm in the gymnosperms, secondary endosperm in the angiosperms), and a protective coating (integument) that may be contributed to by tissues of both the female gametophyte and the sporophyte of the previous generation. Therefore, tissues of three different generations are usually found in a seed: two successive sporophytic generations and the gametophytic generation between them.

At maturity, the seed enters a phase of quiescence or dormancy during which its metabolic functions are largely suspended, to be resumed if and when conditions for germination will occur. There are huge differences between species both in the minimum time of dormancy and in the duration of potential viability, or germinability (from a few weeks to several years). Generally, germinability declines rapidly after the first year, or after a few years, but individual seeds can survive much longer than the average for their species. The length of time during which a seed can remain viable and capable of germinating depends not only on the species, but also, to a considerable extent, on the conditions to which it is exposed. The oldest seed in which germinability has been demonstrated belongs to a small plant of the pink family, the narrow-leafed campion (*Silene stenophylla*), with an age of about 32,000 years for a bunch of seeds that were stored by a squirrel and remained protected under the permafrost, the permanently frozen ground of the Artic regions. Date palm seeds, 2,000 years old and also still able to germinate, were found during archaeological excavations in the palace of Herod the Great at Masada, Israel. Viable lotus seeds were found that had survived 1,300 years in the dried bed of a lake in northeastern China.

The way in which the nourishment provided by the parent is stored in the young embryo may vary from group to group. Among the angiosperms, in most eudicots, one of the two main flowering plant divisions, the cotyledons or embryonic leaves, usually two, accumulate the nutrients that will be used during and after germination. During embryonic development, the cotyledons are filled with starch, oils and proteins, while the endosperm, the source of these nutrients, is reduced until exhaustion. By contrast, in monocots, the other major flowering plant group, the single cotyledon generally remains thin, and the endosperm is still present in the mature seed. During germination, the cotyledon, now called a haustorium, transfers to the embryo the nutritional substances stored in the endosperm, which are progressively mobilized by the action of hydrolytic enzymes the haustorium itself synthesizes. A huge variety of intermediate modes exists between these two extreme modes of embryo supply through the cotyledons.

In angiosperms, seed weight varies over 10 orders of magnitude, from the minute 1-microgram seeds of some orchids to the seeds of the sea coconut, an endemic palm tree from the Seychelles, which can weigh more than 18 kilograms with a diameter of 50 centimetres. As in the common coconut, most of the mass of this gigantic seed (the 'milk' and the 'pulp') consists of triploid endosperm.

The *fruit* is a structure that protects the seeds of the flowering plants and usually contributes to their dispersal. Thus, although the fruit does not provide nourishment to the embryo, it plays a major role in enhancing the potential for dispersal and the chances of survival of the seeds. For this reason, the processes of development that lead to fruit production by the mother plant, and the investment of energy in it, can be compared with parental care in animals. From the point of view of development, the fruit derives from ovarian tissues and is therefore, strictly speaking, a structure exclusive to the flowering plants. However, in some gymnosperms (e.g. in the ginkgo, and in yews and junipers) a fleshy envelope develops around the seed. This cannot strictly be called a fruit, given its different developmental origin, but from the functional point of view it assumes a role similar to that of the fruits of the flowering plants.

In the angiosperms, the fruit may derive exclusively from ovarian tissues (*true fruit*, e.g. the cherry), but its formation may additionally involve tissues of the receptacle, stamens, sepals or petals. In this case it is called an *accessory fruit* or *false fruit*, like the pear, a large part of which develops from the receptacle. A fruit derived from a single carpel or from several carpels fused together is called a *simple fruit*, like the bean, while a single fruit deriving from separate units corresponding to as many carpels is described as an *aggregate fruit* or compound fruit, like the blackberry. Finally, if a single accessory fruit develops from an inflorescence, this is a *composite fruit* or multiple fruit, like the pineapple and the fig. Fruits are also classified into dry and fleshy. Some dry fruits open at maturity, allowing the dispersal of seeds (*dehiscent dry fruits*), while others remain closed (*indehiscent dry fruits*). Fleshy fruits are generally destined for consumption by animals, which contribute in various ways to their dissemination. Most fleshy fruits are indehiscent, but there are a few interesting exceptions. The cells of the tissues around the seeds in the ripe fruit of the squirting cucumber (*Ecballium elaterium*) turn into a mucilaginous liquid that is easily expelled, together with the seeds, due to the high osmotic pressure (up to 27 atmospheres) created by the high concentration of the cytoplasm. In such a state, a light touch may cause the fruit to squirt vigorously, dispersing the seeds all around. A different mechanism is involved in the sudden dehiscence of the fruits of the many species of *Impatiens*, also known as touch-me-not. In this case, a slight bump triggers the dissociation of the fruit into five valves, each of which wraps around itself in a spiral, allowing the seeds to run out. Explosion at maturity also characterizes a few dry fruits, and in this respect those of the sandbox tree (*Hura crepitans*) are spectacular. Fruits are large capsules, up to

8 centimetres in diameter, which at maturity dissociate into 16 segments, launching the seeds a distance of up to 50 metres at a speed of 70 metres per second. No wonder this tree is also known as the dynamite tree.

Similar to pollen dispersal, fruit (and seed) dispersal can rely on different agents: wind (*anemochory*; e.g. poplar, maple and dandelion), water (*hydrochory*; e.g. sea rocket and water lily), or animals (*zoochory*). The fruits or seeds of a number of plants travel attached to the animal thanks to anchoring devices, such as the hooks of many spiny fruits (e.g. spiny cocklebur); others, including the coloured fruits with fleshy pulp that so many birds are fond of, are eaten by the animal, which will release the seeds upon defecation.

There is a huge variation in the number of fruits produced by an individual plant (generally high in the case of small dry fruits, low for large fleshy fruits), energy reserves for the development of each fruit (generally low for dry fruits, higher for fleshy ones), and the number of seeds released in a single fruit (see *Fecundity*, above).

So here we are. It has been a long journey through the intricacies of sex and reproduction in living beings. Well equipped with the knowledge that we have gained from all that we have seen, we should now dare to abandon the comfort zone of the seashore, venturing into the deep waters of some difficult questions. We will engage with a number of these in the Concluding Remarks that follow.

Concluding Remarks: Difficult Boundaries

Setting Sail

Delimiting the biological process we call reproduction with respect to other biological processes is not as easy as it may first appear. Exploring these 'difficult boundaries' – or, if you prefer, these 'grey areas' – is not a mere academic exercise. Our descriptions and classifications can significantly affect investigations and understanding, allowing similarities and differences among the reproductive phenomena to appear, or hiding them from our sight. Our schemes may prove adequate to address some questions, while at the same time being totally inadequate to look into others. As we approach the end of the book, we are now equipped to look at some of these difficult boundaries.

Reproduction or Development?

The typical life cycle of cnidarians is described as a multigenerational cycle: the transition from polyp to medusa is interpreted as a reproductive event separating two distinct generations, in an alternation called *metagenesis*. The typical life cycle of sea urchins and sea stars is described instead as a monogenerational cycle: the transition from larva to adult (or juvenile) is interpreted as a *metamorphosis*, in other words as the transformation of one and the same individual, with no reproductive event involved. But is the distinction between metagenesis and metamorphosis always so clear?

A first criterion, based on demography, would see reproduction when there is an increase in the number of individuals. In most cnidarians the detachment of one or more medusae from the parent polyp does indeed lead to an increase in the number of individuals. But what if the polyp disappears in giving life to a single medusa? In cubozoan cnidarians, one polyp 'transforms' directly into one medusa. This seems to be a developmental process, rather than a reproductive event. Should we call the cubozoan polyp a larva and claim that the cycle of these cnidarians is monogenerational?

If the demographic criterion does not clearly separate metagenesis from metamorphosis, we could perhaps focus on the fact that in asexual reproduction by budding the parent survives after the detachment of its descendants, while nothing survives metamorphosis except for the metamorphosed organism. But, again, in the case of the cubozoans, no surviving polyp accompanies the freshly formed medusa, and in the hydrozoan *Eirene hexanemalis* the polyp is even planktonic (like a sea urchin larva) and produces by budding a single medusa that completely reabsorbs what remains of the polyp. This cycle can be clearly described as monogenerational, exactly as in sea urchins.

Possibly, another divide between metagenesis from metamorphosis might be found in the details of the two processes, arguing that in metagenesis reproduction occurs through buds that are only a part of the individual parent, while metamorphosis is a transformation of an entire individual. However, in the metamorphosis of many forms of marine invertebrates, most of the larval body is discarded or consumed and the young derives from a small number of founding cells. In the sea star *Luidia sarsii* the larva can even continue to swim three months after the juvenile that originated from it has detached. Should we say that (what remains of) the larva of this echinoderm reproduces asexually, and that it has a metagenetic life cycle involving two generations, like a cnidarian's?

Actually, whenever evolution has preserved both polyp and medusa, cnidarian cycles are invariably described in terms of metagenesis, regardless of how the polyp-to-medusa transition occurs. In parallel, sea urchin and sea star life cycles that involve a larval phase are invariably described in terms of metamorphosis, regardless of how the larva-to-adult transition occurs. In many cases, a distinction between reproduction and metamorphosis is reduced to a lexical question, or to a question of taxon-specific traditions.

Reproduction or Growth?

If one considers asexual reproduction as nothing but a form of growth (with fragmentation) of the individual, as some authors argue, a sequence of asexual generations would reduce to a process of expansion and transformation of the body (not necessarily cohesive) of a single individual.

If we stipulate that genetic identity is the fundamental criterion defining a biological individual (Chapter 1), any form of clonal propagation that produces multiple copies of an individual's genotype should not be seen as reproduction, but rather as the growth of one individual. What counts as reproduction can be a matter of perspective, depending on how an individual is defined.

Applying the criterion of genetic identity, all the polyps of a coral would not be members of a colony of individuals, but rather parts of the same (super)organismal individual. In all cnidarians there would not be cycles with alternation of sexual and asexual generations (metagenesis): these would be reduced to monogenerational cycles, with a metamorphosis followed by a particular kind of growth. Even more strangely, a lawn of dandelions, a plant that propagates by apomixis (Chapter 6), could be seen as a 'large diffuse tree' that has not invested energy and materials building a woody trunk, branches and a persistent root system. The entire lawn would be a huge, although scattered, 'genetic individual', consisting of many physically separate parts, each having the capacity to grow further and possibly to reproduce.

An emblematic case of asexual reproduction, or from a different perspective, of individual growth, is a 'vegetal entity' known as Pando. What looks like a forest of quaking aspen covering about 45 hectares in Utah is nothing but a clone that would constitute, according to some, a single (male) living organism, which weighs about 6,600 tonnes and includes some 47,000 trunks that continually decay and are regenerated by a single, gigantic root system. This organism would be about 80,000 years old and would thus be the heaviest and oldest living organism known. However, since Pando's root system has probably fragmented over time into a set of contiguous but disconnected subsystems, should it still be considered a single individual? Moreover, since the somatic mutations accumulated over such a long period of time make it genetically very heterogeneous, to levels typical of a population of individuals, should we still regard it as a single individual?

Reproduction or Regeneration?

Regeneration is a developmental process that allows an individual to replace a lost part of its body. There is surely no danger of mistaking regeneration for reproduction, or vice versa. However, the distinction is not always so clear-cut, particularly for those organisms capable of complete regeneration (whole-body regeneration), as seen, among animals, in many cnidarians, annelids and flatworms. In the most common forms of asexual reproduction of multicellular organisms, a part of the parent's body differentiates into what will become the propagule (a hydra bud, a lily bulbil, the posterior terminal section of the trunk of some marine worms), founder of an individual of the offspring generation that eventually detaches from the parent to further develop and lead independent life. However, in asexual reproduction by architomy (Chapter 3), first a small piece of the individual parent detaches, carrying the original tissue organization, and only after detachment this piece (re)generates all that is missing, to create another complete independent individual. This is what happens for example in the freshwater annelid *Lumbriculus*. It seems difficult to establish here a clear boundary between asexual reproduction by fragmentation and regeneration.

In the development of a part of an individual's body that will become a new individual through asexual reproduction, at which point do we place the transition from 'part of the parent's body' to 'new individual offspring'? The problem of regeneration adds to this problem a further reason for uncertainty. Should the answer depend on whether those cells are already committed to becoming another individual? In fact, while the formation of a bud in a hydra anticipates its future detachment as an autonomous individual, there is nothing in the development of a portion of the body of certain annelids that qualifies it as a part destined for reproduction, although it may contribute to reproduction following accidental separation. However, on closer examination, it is observed that both regeneration and asexual reproduction depend on the availability of undifferentiated cells (or cells that can return to an undifferentiated state after differentiation) ready to multiply to rebuild a complete body through appropriate morphogenetic processes. There seem to be no factors that limit the activity of these cells to the exclusive service of one or the other process.

Again, What Is Reproduction?

In Chapter 1 we answered this question pragmatically, but what can we say now, at the end of our excursion through the diversity of reproductive phenomena and their interconnections? We should have developed a richer perspective on the subject, but at the same time an awareness of the difficulties involved in tracing neat boundaries that separate reproduction from other biological processes, and in applying rigid classifications.

There are concepts and phenomena which by their very nature do not lend themselves to being neatly circumscribed and defined. Reproduction is possibly one of them. Any attempt to apply arbitrary delimitations and inflexible classifications would easily alienate us from the nature of reproduction. As an alternative, we may follow the suggestion of Ludwig Wittgenstein, the philosopher who in his *Philosophical Investigations*, published in 1953, discusses similar problems with definitions. Reproduction would thus be 'a family of related concepts' that revolve around the intuitive and familiar idea of reproduction and, in one sense or another, involve the generation of the living.

A realistically nuanced concept of reproduction will allow us to better explore and appreciate the enormous variety of expedients through which, day after day, living beings achieve continuity through time.

Summary of Common Misunderstandings

This short section recaps the most common unwarranted generalizations in the biology of reproduction. These result from a view of reproduction and reproductive processes biased towards the behaviour of organisms (especially animals) more familiar to us.

Sex and reproduction are entangled biological process. No, they are potentially independent. There can be sex without reproduction, as in the genetic exchange of ciliate conjugation, and reproduction without sex, as in the binary fission of an individual amoeba (Chapter 1).

There are sexual and asexual organisms. No, sexual or asexual are attributes of reproduction, not of an organism. Many organisms can reproduce both ways, sexually (involving sex processes) and asexually (without involving sex processes) (Chapter 1).

Offspring are the same kind of organism as their parents. Not always. There are life cycles that includes more than one reproductive phase carried out by different organizational forms of the same organism. In these cycles with alternation of generations, offspring can differ from parents, as in the sporophyte and the gametophyte of plants (Chapter 2).

Reproduction can be summarized as 'making copies of themselves'. This is not true in general. In multigenerational cycles, offspring are not of the same kind as their parents (Chapter 2).

Gametes are the direct product of meiosis. Not always. Haplontic multicellular organisms produce gametes by mitosis, and haplontic unicellular organisms can transform into a gamete (Chapter 2).

Development starts from a zygote. Not always. It can start also from a spore or from a multicellular propagule, such as a bud (Chapters 2 and 3).

Asexual reproduction produces perfectly clonal descendants. Not necessarily. Genetic mutations and/or phenomena of genetic recombination undermine the perfect genetic uniformity of a multicellular organism, or of the lineage of a unicellular organism (Chapter 3).

Reproduction is an adult affair. Not exclusively. Polyembryony (if viewed as embryonic asexual reproduction), larval amplification, Russian-doll parthenogenesis and paedogenesis are all forms of reproduction during an early life stage (Chapters 3 and 6).

Sexual reproduction results in genetically distinct offspring. Not always. Polyembryony, some forms of parthenogenesis, or even iterated selfing can produce genetically identical individuals (Chapters 3 and 6).

Individuals are either male or female. This is not always the case. In addition to males and females there are the hermaphrodites. These occur exclusively in certain species, while in others they are found together with males, or with females, or with both. In species that reproduce sexually through isogametes, individuals are sexually indeterminate (Chapter 4).

Gametes are of two kinds. This is untrue. There are male gametes (sperm), female gametes (eggs), but also gametes that are neither male nor female in species that produce just one single type of gametes (isogametes) (Chapter 4).

The sex of an individual is fixed. Not in all species. In the case of sequential hermaphrodites, the sex of an individual changes during its life, or can even switch repeatedly (alternating hermaphroditism) (Chapter 4).

Pollen and embryo sac are the reproductive organs of the sporophyte of a seed plant. Not properly so. They are actually its offspring, belonging to the gametophyte generation (Chapter 4).

Sex roles are the same for all organisms. This is not so. In contrast to the more frequent conditions, there are males that incubate the eggs, females that court males, males that provide parental care, females that defend territory, females that fight for males, males that chose the female to mate with (Chapter 4).

Fertilization needs a partner encounter. Not necessarily. There is no mating in organisms where male gametes, or both male and female gametes, are released into the water (free spawning and broadcast spawning, respectively), where fertilization takes place (external fertilization) (Chapter 5).

Pollination is fertilization in plants. Not really. It can actually be viewed as a sexual-partner encounter, that of two gametophytes. In some conifers more than one year can pass between pollination (pollen arrival in proximity of the female gametophyte) and fertilization (syngamy) (Chapter 5).

Incest is avoided in nature. This is not a rule. Animal avoidance of mating with relatives is not a must, and incestuous mating is regular or even obligate in some species. For instance, oedipal (mother–son) mating is obligate in the mite *Histiostoma murchiei*, while brother–sister mating is obligate in another mite, *Adactylidium* (Chapter 5).

There are always two parents. This is untrue. Asexual reproduction, parthenogenesis and self-fertilization are common forms of reproduction from a single parent (uniparental reproduction). The number of parents can also be somewhere between one and two, as in gynogenesis and other unusual modes of sexual reproduction (Chapter 6).

An individual's sex (or sex condition) is a chromosomal affair. Not in all species. There are many forms of determination of sex (or sex condition) where the genotype of the individual is irrelevant, as in the various forms of environmental and maternal sex determination (Chapter 7).

Genetic sex-determination systems involve sex chromosomes. Not necessarily. There are genetic systems of sex determination where sex-determining genes or alleles are sparse across the whole genome (Chapter 7).

When parental care is provided by a single parent, this is always the mother. Not in all species. In seahorses and sea spiders, and in certain birds and frogs, it is the male who incubates the eggs (Chapter 8).

Family is a safe haven. This is fiction. There are mothers devoured by their offspring (extreme suicidal matrotrophy, as in the spider *Stegodyphus lineatus*), males eaten up by their female partner (mantis), offspring generated only to nourish their brothers and sisters (oophagy in ants and adelphophagy in salamanders) (Chapter 8).

A seed is simply a young plant. No, it is actually a complex structure, which includes tissues belonging to three different generations: two successive sporophytic generations and the gametophytic generation between them (Chapter 8).

References and Further Reading

As explained in the Preface, this book draws extensively from our book *The Biology of Reproduction* (2019). In that graduate-level textbook, all the topics discussed here are treated in greater depth, with more detailed reference to the original literature and with more extensive and precise reports on the taxonomic distribution of the different modes of reproduction. Thus, for each chapter of this book, you can assume an implicit reference to our earlier volume. In the following list, we include references to important books or textbooks on the subject, and to review papers or other research papers that in their introduction review the state of the art of present knowledge. Some titles, although listed under a specific chapter, may be relevant to more than one chapter.

Chapter 1

Agate, R. J., Grisham, W., Wade, J., Mann, S., Wingfield, J., Schanen, C., Palotie, A., and Arnold, A. P. 2003. Neural, not gonadal, origin of brain sex differences in a gynandromorphic finch. *Proceedings of the National Academy of Sciences of the United States of America* 100: 4873–4878.

Alberts, B., Heald, R., Johnson, A. D., Morgan, D., and Raff, M. (2022). *Molecular Biology of the Cell*, 7th edition. New York: Norton.

Bell, G. (1988). *Sex and Death in Protozoa: The History of an Obsession*. Cambridge: Cambridge University Press.

Hanschen, E. R., Davison, D. R., Grochau-Wright, Z. I., and Michod, R. E. (2017). Evolution of individuality: a case study in the volvocine green algae. *Philosophy, Theory, and Practice in Biology* 9: 3.

Jones, O. R., Scheuerlein, A., Salguero-Gómez, R., Camarda, C. G., Schaible, R., Casper, B. B., Dahlgren, J. P., Ehrlén, J., García, M. B., Menges, E.S., Quintana-Ascencio, P. F., Caswell, H., Baudisch, A., and Vaupel, J. W. (2014). Diversity of ageing across the tree of life. *Nature* 505: 169–173.

Krebs, J. E., Goldstein, E. S., and Kilpatrick, S. T. (2017). *Lewin's Genes XII*. Burlington, MA: Jones & Bartlett Learning.

Maynard Smith, J., and Szathmáry, E. (1995). *The Major Transitions in Evolution*. San Francisco, CA: Freeman.

Rovelli, C. (2017). *L'ordine del tempo*. Milan: Adelphi. Published in English as *The Order of Time*. London: Allen Lane, 2018.

Santelices, B. (1999). How many kinds of individual are there? *Trends in Ecology and Evolution* 14: 152–155.

Sender, R., and Milo, R. (2021). The distribution of cellular turnover in the human body. *Nature Medicine* 27: 45–48.

Shefferson, R. P., Jones, O. R., and Salguero-Gómez, R. (2017). *The Evolution of Senescence in the Tree of Life*. Cambridge: Cambridge University Press.

Sheldrake, A. R. (2022). Cellular senescence, rejuvenation and potential immortality. *Proceedings of the Royal Society B* 289: 20212434.

Susskind, L., and Hrabovsky, G. (2013). *The Theoretical Minimum: What You Need to Know to Start Doing Physics*. New York: Basic Books.

Chapter 2

Bowman, J. L., Sakakibara, K., Furumizu, C., and Dierschke, T. (2016). Evolution in the cycles of life. *Annual Review of Genetics* 50: 133–154.

Cross, F. R., and Umen, J. G. (2015). The *Chlamydomonas* cell cycle. *The Plant Journal* 82: 370–392.

Fusco, G. (2019). Evo-devo beyond development: the evolution of life cycles. In G. Fusco (ed.), *Perspectives on Evolutionary and Developmental Biology*. Padova: Padova University Press, pp. 309–318.

Haufler, C. H., Pryer, K. M., Schuettpelz, E., Sessa, E. B., Farrar, D. R., Robbin Moran, J., Schneller, J., Watkins, J. E., and Windham, M. D. (2016). Sex and the single gametophyte: revising the homosporous vascular plant life cycle in light of contemporary research. *BioScience* 66: 928–937.

Heesch, S., Serrano-Serrano, M., Barrera-Redondo, J., Luthringer, R., Peters, A. F., Destombe, C., Cock, J. M., Valero, M., Roze, D., Salamin, N., and Coelho S. M. (2021). Evolution of life cycles and reproductive traits: Insights from the brown algae. *Journal of Evolutionary Biology* 34: 992–1009.

Heming, B. S. (2003). *Insect Development and Evolution*. Ithaca, NY: Comstock.

Otto, S. P., and Gerstein, A. C. (2008). The evolution of haploidy and diploidy. *Current Biology* 18: R1121–R1124.

Sorojsrisom, E. S., Haller, B. C., Ambrose, B. A., and Eaton D. A. R. (2022) Selection on the gametophyte: modeling alternation of generations in plants. *Applications in Plant Sciences* 10: e11472.

Chapter 3

Avise, J. C. (2008). *Clonality: The Genetics, Ecology and Evolution of Sexual Abstinence in Vertebrate Animals*. New York: Oxford University Press.

Bell, A. D. (2008). *Plant Form: an Illustrated Guide to Flowering Plant Morphology*, 2nd edition. Portland, OR: Timber Press.

Brusca, R. C., Giribet, G., and Moore, W. (2022). *Invertebrates*, 4th edition. Sunderland, MA: Sinauer Associates.

Lushai, G., and Loxdale, H. D. (2002). The biological improbability of a clone. *Genetics Research* 79: 1–9.

Lynch, M. (2010). Evolution of the mutation rate. *Trends in Genetics* 26: 345–352.

Quiroga, H. (1921). Gloria tropical. In *Anaconda*. Buenos Aires: Agencia Gral. de Librería y Publicaciones.

Chapter 4

Aanen, D., Beekman, M., and Kokko, H. (eds.) (2016). Weird sex: the underappreciated diversity of sexual reproduction. *Philosophical Transactions of the Royal Society B* 371: 20160262.

Avise, J. C. (2011). *Hermaphroditism: A Primer on the Biology, Ecology, and Evolution of Dual Sexuality*. New York: Columbia University Press.

Barrett, S. C. (2002). The evolution of plant sexual diversity. *Nature Reviews Genetics* 3: 274–284.

Bell, G. (1982). *The Masterpiece of Nature: the Evolution and Genetics of Sexuality*. London: Croom Helm.

Burke, N. W., and Bonduriansky, R. (2017). Sexual conflict, facultative asexuality, and the true paradox of sex. *Trends in Ecology and Evolution* 32: 646–652.

Johnson, G. D., Paxton, J. R., Sutton, T. T., Satoh, T. P., Sado, T., Nishida, M., and Miya, M. (2009). Deep-sea mystery solved: astonishing larval transformations and extreme sexual dimorphism unite three fish families. *Biology Letters* 5: 235–239.

Leonard, J. L. (2018). The evolution of sexual systems in animals. In J. L. Leonard (ed.), *Transitions Between Sexual Systems*. Cham: Springer, pp. 1–58.

Lloyd, D. G., and Webb, C. J. (1977). Secondary sex characters in plants. *The Botanical Review* 43: 177–216.

Ni, M., Feretzaki, M., Sun, S., Wang, X., and Heitman, J. (2011). Sex in fungi. *Annual Review of Genetics* 45: 405–430.

Otto, S. P. (2009). The evolutionary enigma of sex. *American Naturalist* 174 Suppl 1: S1–S14.

Whittle, C. A., and Extavour, C. G. (2017). Causes and evolutionary consequences of primordial germ-cell specification mode in metazoans. *Proceedings of the National Academy of Sciences of the United States of America* 114: 5784–5791.

Chapter 5

Alcock, J. (2013). *Animal Behavior: an Evolutionary Approach*, 10th edition. Sunderland, MA: Sinauer.

Billiard, S., López-Villavicencio, M., Devier, B., Hood, M. E., Fairhead, C., and Giraud, T. (2011). Having sex, yes, but with whom? Inferences from fungi on the evolution of anisogamy and mating types. *Biological Reviews* 86: 421–442.

de Boer, R.A., Vega-Trejo, R., Kotrschal, A., and Fitzpatrick, J. L. (2021) Meta-analytic evidence that animals rarely avoid inbreeding. *Nature Ecology & Evolution* 5: 949–964.

Egevang, C., Stenhouse, I. J., Phillips, R. A., Petersen, A., Fox, J. W., and Silk, J. R. D. (2010). Tracking of Arctic terns *Sterna paradisaea* reveals longest animal migration. *Proceedings of the National Academy of Sciences of the United States of America* 107, 2078–2081.

Fernando, D. D., Lazzaro, M. D., and Owens, J. N. (2005). Growth and development of conifer pollen tubes. *Sexual Plant Reproduction* 18: 149–162.

Harada, Y., Takagaki, M., Sunagawa, M., Saito, T., Yamada, L., Taniguchi, H., Shoguchi, E., and Sawada, H. (2008). Mechanism of self-sterility in a hermaphroditic chordate. *Science* 320: 548–550.

Harder, L. D., and Barrett, S. C. H. (2006). *Ecology and Evolution of Flowers*. New York: Oxford University Press.

Mauseth, J. (2014). *Botany: an Introduction to Plant Biology*, 4th edition. Burlington, MA: Jones & Bartlett.

Oliveira, R. F., Taborsky, M., and Brockmann, H. J. (eds.) (2008). *Alternative Reproductive Tactics: An Integrative Approach*. Cambridge: Cambridge University Press.

Pesendorfer, M. B., Ascoli, D., Bogdziewicz, M., Hacket-Pain, A., Pearse, I.S., and Vacchiano, G. (2021). The ecology and evolution of synchronized reproduction in long-lived plants. *Philosophical Transactions of the Royal Society B: Biological Sciences* 376: 20200369.

Richards, A. J. (1997). *Plant Breeding Systems*, 2nd edition. London: Chapman & Hall.

Rosenstiel, T. N., Shortlidge, E. E., Melnychenko, A. N., Pankow, J. F., and Eppley, S. M. (2012). Sex-specific volatile compounds influence microarthropod-mediated fertilization of moss. *Nature* 489: 431–433.

Westneat, D., and Foz, C. (eds.) (2010). *Evolutionary Behavioral Ecology*. Oxford: Oxford University Press.

Chapter 6

Beukeboom, L. W., and Vrijenhoek, R. C. (1998). Evolutionary genetics and ecology of sperm-dependent parthenogenesis. *Journal of Evolutionary Biology* 11: 755–782

Gokhman, V. E., and Kuznetsova, V. G. (2018). Parthenogenesis in Hexapoda: holometabolous insects. *Journal of Zoological Systematics and Evolutionary Research* 56: 23–34.

Jarne, P., and Auld, J. R. (2006). Animals mix it up too: the distribution of self-fertilization among hermaphroditic animals. *Evolution* 60: 1816–1824.

Mogie, M. (1992). *The Evolution of Asexual Reproduction in Plants*. London: Chapman & Hall.

Neaves, W. B., and Baumann, P. (2011). Unisexual reproduction among vertebrates. *Trends in Genetics* 27: 81–88.

Schön, I., Martens, K., and van Dijk, P. (eds.) (2009). *Lost Sex: The Evolutionary Biology of Parthenogenesis*. Berlin: Springer.

Vershinina, A. O., and Kuznetsova, V. G. (2016). Parthenogenesis in Hexapoda: Entognatha and non-holometabolous insects. *Journal of Zoological Systematics and Evolutionary Research* 54: 257–268.

Chapter 7

Barresi, J. F., and Gilbert, S. F. (2020). *Developmental Biology*, 12th edition. New York: Oxford University Press.

Beukeboom, L. W., and Perrin, N. (2014). *The Evolution of Sex Determination*. Oxford: Oxford University Press.

Fusco, G., and Minelli, A. (2023). Descriptive versus causal morphology: gynandromorphism and intersexuality. *Theory in Biosciences* 142: 1–11.

Hardy, I. C. W. (ed.) (2002). *Sex Ratios: Concepts and Research Methods*, Cambridge: Cambridge University Press.

Janousek, B., and Mrackova, M. (2010). Sex chromosomes and sex determination pathway dynamics in plant and animal models. *Biological Journal of the Linnean Society* 100: 737–752.

Kaiser, V. B., and Bachtrog, D. (2010). Evolution of sex chromosomes in insects. *Annual Review of Genetics* 44: 91–112.

Katona, G., Vági, B., Végvári, Z., Liker, A., Freckleton, R. P., Bókony, V., and Székely. T. (2021). Are evolutionary transitions in sexual size dimorphism related to sex determination in reptiles? *Journal of Evolutionary Biology* 34: 594–603.

Love, A. C., and Yoshida, Y. (2019). Reflections on model organisms in evolutionary developmental biology. In W. Tworzydlo and S. M. Bilinkski (eds.), *Evo-Devo: Non-Model Species in Cell and Developmental Biology*. Cham: Springer, pp. 3–20.

Ming, R., Bendahmane, A., and Renner, S. S. (2011). Sex chromosomes in land plants. *Annual Review of Plant Biology* 62: 485–514.

Sánchez, L. (2008). Sex-determining mechanisms in insects. *International Journal of Developmental Biology* 52: 837–856.

White, M. J. D. (1973). *Animal Cytology and Evolution*. Cambridge: Cambridge University Press.

Chapter 8

Althaus, S., Jacob, A., Graber, W., Hofer, D., Nentwig, W., and Kropf, C. (2010). A double role of sperm in scorpions: the copulatory plug of *Euscorpius italicus* (Scorpiones: Euscorpiidae) consists of sperm. *Journal of Morphology* 271: 383–393.

Behie, S. W., and Bidochka, M. J. (2014). Nutrient transfer in plant–fungal symbioses. *Trends in Plant Science* 19: 734–740.

Bewley, J. D., Bradford, K. J., Hilhorst, H. W. M., and Nonogaki, H. (2013). *Seeds: Physiology of Development, Germination and Dormancy*. New York: Springer.

Cieslak, A., Fresneda, J., and Ribera, I. (2014). Life-history specialization was not an evolutionary dead-end in Pyrenean cave beetles. *Proceedings of the Royal Society B* 281: 20132978.

Garbiec, A., Christophoryová, J., and Jędrzejowska, I. (2022). Spectacular alterations in the female reproductive system during the ovarian cycle and adaptations for matrotrophy in chernetid pseudoscorpions. *Scientific Reports* 12: 6447.

Hughes, P. W. (2017). Between semelparity and iteroparity: empirical evidence for a continuum of modes of parity. *Ecology and Evolution* 7: 8232–8261.

Nath, P., Bouzayen, M., Mattoo, A. K., and Pech, J. C. (eds.) (2014). *Fruit Ripening: Physiology, Signalling and Genomics*. Wallingford: CABI.

Ostrovsky, A., Lidgard, S. Gordon, D., Schwaha, T., Genikhovich, G., and Ereskovsky, A. (2016). Matrotrophy and placentation in invertebrates: a new paradigm. *Biological Reviews* 91: 673–711.

Poethig, R. S. (2003). Phase change and the regulation of developmental timing in plants. *Science* 301: 334–336.

Price, C. S. C., Dyer, K. A., and Coyne, J. A. (1999). Sperm competition between *Drosophila* males involves both displacement and incapacitation. *Nature* 400: 449–452.

Vogt, G. (2016). Structural specialties, curiosities and record-breaking features of crustacean reproduction. *Journal of Morphology* 277: 1399–1422.

Concluding Remarks

Fusco, G., and Minelli, A. (2019). *The Biology of Reproduction*. Cambridge: Cambridge University Press.

Godfrey-Smith, P. (2009). *Darwinian Populations and Natural Selection*. New York: Oxford University Press.

Janzen, D. H. (1977). What are dandelions and aphids? *American Naturalist* 111: 586–589.

Salazar-Ciudad, I. (2006). Evolution in biological and non-biological systems under different mechanisms of generation and inheritance. *Theory in Biosciences* 127: 343–358.

Figure Credits

All the figures are originals, produced specifically for this book by Mariagiulia Sottoriva. Figures derive from various sources, or are redrawn by hand from those sources, as follows:

Figures 1.1, 6.1, 7.2 and 8.1 from Fusco, G., and Minelli, A. (2019). *The Biology of Reproduction*. Cambridge: Cambridge University Press.

Figure 1.2 from Rovelli, C. (2017). *L'ordine del tempo*. Milan: Adelphi.

Figure 1.3 from Henderson, K. A., and Gottschling, D. E. (2008). A mother's sacrifice: what is she keeping for herself? *Current Opinion in Cell Biology* 20: 723–728.

Figure 2.1 from Wangler, M. F., and Bellen, H. J. (2017). In vivo animal modeling: *Drosophila*. In Jalali, M., Saldanha, F. Y. L., and Jalali, M. *Basic Science Methods for Clinical Researchers*. London: Academic Press, pp. 212–235.

Figures 2.2, 2.3 and 4.1 from Mauseth, J. D. (2009). *Botany: an Introduction to Plant Biology*, 4th edition. Burlington, MA: Jones & Bartlett.

Figure 2.5 from Bayer, F. M., and Owre, H. B. (1968). *The Free-Living Lower Invertebrates*. London: Macmillan.

Figure 3.2 from Zattara, E. E., and Bely, A. E. (2016). Phylogenetic distribution of regeneration and asexual reproduction in Annelida: regeneration is ancestral and fission evolves in regenerative clades. *Invertebrate Biology* 135: 400–414.

Figure 5.1 from Silvertown, J., and Charlesworth, D. (2001). *Introduction to Plant Population Biology*, 4th edition. Oxford: Blackwell.

Figure 7.1 from Bachtrog, D., Kirkpatrick, M., Mank, J. E., *et al.* (2011). Are all sex chromosomes created equal? *Trends in Genetics* 27: 350–357.

Index

Locators in **bold** refer to tables, those in *italic* to figures

Printed in the United States
by Baker & Taylor Publisher Services

Printed in the United States
by Baker & Taylor Publisher Services